DAYSTAR

A Peep into the Workings of the Sun

DAYSTAR

A Peep into the Workings of the Sun

Parameswaran Venkatakrishnan

Indian Institute of Astrophysics, India

World Scientific

NEW JERSEY · LONDON · SINGAPORE · BEIJING · SHANGHAI · HONG KONG · TAIPEI · CHENNAI · TOKYO

Published by

World Scientific Publishing Co. Pte. Ltd.

5 Toh Tuck Link, Singapore 596224

USA office: 27 Warren Street, Suite 401-402, Hackensack, NJ 07601

UK office: 57 Shelton Street, Covent Garden, London WC2H 9HE

National Library Board, Singapore Cataloguing in Publication Data

Names: Venkatakrishnan, P., author.

Title: Daystar : a peep into the workings of the Sun / Parameswaran
Venkatakrishnan (Indian Institute of Astrophysics, India).

Description: Singapore ; Hackensack, NJ : World Scientific, [2018] | Includes
bibliographical references and index.

Identifiers: LCCN 2017033920| ISBN 9789813228528 (hardcover ; alk. paper) |
ISBN 9813228520 (hardcover ; alk. paper)

Subjects: LCSH: Sun.

Classification: LCC QB521 .V46 2018 | DDC 523.7--dc23

LC record available at https://lccn.loc.gov/2017033920

British Library Cataloguing-in-Publication Data

A catalogue record for this book is available from the British Library.

Dedicated to the memory of

my parents:

R. Parameswara Iyer (1919–1965)
&
Meenakshi Parameswaran (1929–2016)

and also to my mentor:

Professor U.R. Rao (1932–2017)

भारतीय अन्तरिक्ष अनुसंधान संगठन
अन्तरिक्ष विभाग, भारत सरकार, अन्तरिक्ष भवन
न्यू बी ई एल रोड, बंगलूरु – 560 231, भारत
दूरभाष : +91 80 2341 6406 / 2217 2387
फैक्स : +91 80 2341 0705

Indian Space Research Organisation
Department of Space, Government of India
Antariksh Bhavan, New BEL Road, Bangalore - 560 231, India
Tel (Off) : +91 80 2341 6406 / 2217 2387
Fax   : +91 80 2341 0705
e-mail : urrao.isro@gmail.com

Prof. U R Rao
(Former Chairman, Space Commission &
Secretary, Department of Space)
Chairman, Karnataka Science & Technology Academy
Chancellor, Indian Institute of Space Science and Technology

May 17, 2017

FOREWORD

From the very beginning of human civilization, we have looked at the Sun with great awe and reverence for its ability to bring light into our life, and enable humanity as a whole to survive and thrive. Yet, the Sun, in spite of its pivotal role in life, continues remaining a mystery yet to be fully uncovered. For example, some of the problems, like that of the million-degree corona, have remained unsolved to this day in spite of many space missions from many different countries. The Indian Space Research Organisation has also joined in this global effort by approving its space mission Aditya-L1 to put an observatory class satellite at the L1 Lagrangian point to continuously watch the solar corona. In this context, there is a great need for the influx of the brightest young minds to use the results of this mission and help solve the problem of coronal heating. In spite of familiarizing ourselves with the Sun, a detailed understanding of it and a complete understanding of its effect on the Earth and our life are still elusive.

Dr. Venkatakrishnan, who was the Director of the Udaipur Solar Observatory for over 15 years, is a well-known astronomer. Having carried out original, significant work and investigations on the Sun, he is considered an expert on solar physics. In writing this book he has taken all care to explain our accumulated knowledge of the Sun and has beautifully summarized what we still do not know. The book also covers some very important information on the solar atmosphere and clearly discusses the role of solar activity in climate change.

The entire physics of the Sun is described in six chapters, starting from how the vital statistics of the Sun are measured. Dr. Venkatakrishnan then goes on to explain how the scientists used measurements of the Sun's light to understand what the Sun is made of, how it is put together, how this great mass of mainly hydrogen plasma produces the sunlight which we see, what are the nature and cause of the blemishes which we see on the Sun's face, what eruptions come out of such blemishes, and how these eruptions affect the Earth. Finally, the last chapter describes the different kinds of solar telescopes operating at different bands of the electromagnetic spectrum, including even the latest discoveries about solar neutrinos.

This book on the Sun is indeed a great pleasure to read and provides up-to-date information on the solar interior, atmosphere, magnetic fields, and the effects of solar activity on the Earth, in a very fluent style. This will enable students of solar physics, in the initial stages of their carrier, to easily understand the physics of solar phenomena. I am extremely happy to note that Dr. Venkatakrishnan has brought out not only all the known facts about the Sun in a simple and beautiful style, but also the important unexplained aspects of solar activity. I have no doubt that this book will please a large number of people who are interested in solar physics.

U. R. Rao

PREFACE

While delivering talks about the Sun to schoolchildren, I always found that they would start getting more interested (in the talk) and pay a lot more attention if I appealed to their own knowledge of high school physics. What is really fascinating is to sense their awe when they realize that all the laws of physics applicable to processes in the laboratory on the Earth are applicable anywhere in the Universe. This is not a trivial matter. There is a famous episode in scientific history where Sir Arthur Eddington refused to accept the finding of the young Subrahmanyan Chandrasekhar that some stars can collapse into nothing. This is an important example where people refuse to believe that the laws of physics are rigorously applicable to any situation, no matter where. It is for this reason that I have attempted to demystify the activities of a celestial object such as the Sun in terms of high school physics. To further demystify the methods of obtaining all the facts about the Sun, I have included a chapter on the different kinds of solar telescopes operating at different wavelengths and also at different locations ranging from outer space to deep underground. Happy reading!

P. Venkatakrishnan
Former Head, Udaipur Solar Observatory, Udaipur, India
Honorary Professor, Indian Institute of Astrophysics, Bangalore, India

Bengaluru, India
May 2017

CONTENTS

1 INTRODUCTION

Every morning, the eastern horizon bursts into a colorful song of light, heralding the arrival of our daytime star — the Sun. This glorious event occurs with so much regularity that we tend to pay scant attention to the lord of the day sky. Once in about every generation of men and women, this celestial visitor of regular habits plays a trick on its earthbound spectators, by hiding behind the Moon. It is only then that people become fully aware of the fact that the Sun is no ordinary star. I had the good fortune to witness this spectacular phenomenon of the total eclipse of the Sun four times, and noticed on all the occasions the awe with which people greeted the unveiling of the solar corona — the bluish glittering halo of hot gas surrounding the Sun. What impresses most people is the total unexpectedness of the apparition. Even solar physicists, who know about the existence of the corona from their textbooks, are much unprepared for the actual scene. It is only on such occasions that we are reminded of the exotic physical processes that go on in the Sun. The story of the identification of some of these processes is interesting as well as instructive. At this point in time, the investigations have even laid bare some of the facts in the deep recesses of the Sun. As a by-product, we have also received new knowledge about the very structure of matter itself.

From the dawn of civilization, the Sun has dominated the lives and thoughts of humankind. The earliest of civilized settlements — the Egyptians — conferred a central status on Ra, the sun god. The Aryans made much of the illuminative power of the Sun, leading to daily rituals related to Sun gazing and Sun worship. The Inca and Maya civilizations too had Sun-related rituals. The Greeks had Apollo, the handsome lord of the day sky riding the glorious chariot drawn by fiery horses. For more details, the reader is referred to the illuminating book *Fundamentals of Solar Physics,* by Bhatnagar and Livingston. All this early fuss about the Sun is understandable, considering the fact that the Sun provided the much-needed light to protect early humans from predators that roamed the forests freely in the darkness of the night.

In spite of the present successes of technology, the Sun continues to influence human activities in a variety of ways. Solar energy is the origin of all the sources of energy that we use every day. The coal and petroleum burnt in innumerable engines are relics of solar energy stored in the fossils of plants and animals. The vast hydro-electric power plants essentially tap the potential energy of water that rises from the oceans heated by solar radiation. The modern concepts of renewable energy sources rely on solar energy directly or on the energy of wind, which originates from the thermal gradients in the atmosphere created by solar heating. Finally, the ultimate hope for a large quantity of safe, clean energy is from nuclear fusion, a process that goes on within the center of the Sun. The terrestrial version of this process needs a clever way to contain and heat plasmas, while the Sun effortlessly heats the coronal plasma. We have much to learn from the Sun.

This is not all. Modern life is replete with the conveniences offered by modern communication systems. These systems are maintained using links with a flotilla of satellites revolving around the Earth. Sporadic outbursts of high-energy particles and ejections of magnetized plasma from the Sun can sometimes severely affect these communication systems. These events, called solar flares and coronal mass ejections (CMEs) respectively, can do worse — they can sometimes even trigger large-scale currents in electrical grids (especially at extreme latitudes) that can trip the power systems and cause power outages. Transcontinental air travelers may face radiation hazards while flying over the polar regions during heavy sunstorms. In fact, the havoc wrought by the Carrington flare of 1858 is described very vividly in *The Sun Kings* by Stuart Clark.

For the astrophysicist, the Sun is the only star that can be studied in great detail, purely because it is the nearest star. All the intricate theories about the way in which stars age and die depend on a few physical principles. We can test these theories only in the Sun. Further, only a detailed knowledge of stars can give us an insight into the workings of systems of stars like galaxies, and a grasp of galactic physics is essential for understanding groups of galaxies, which ultimately leads to some understanding of the very universe in which we live. As you can see, the Sun therefore leads to a better understanding of the very small as well as the very large. This dual connection between the microphysics and the macrophysics makes the Sun an extremely fascinating object of study.

Not all this understanding came just for the asking. Even the most fundamental properties of the Sun, viz. its size, mass, and total energy output, could be measured only after the distance from the Earth to the Sun was measured. The story of this measurement merits detailed retelling. The next chapter explains how all this was done. Once we know the fundamental properties of the Sun, especially its energy output, we will face the question of how the Sun generates this much power. To understand this process, we must learn how the Sun holds its matter together, as well as how the fundamental constituents of matter hold themselves together. This will form the subject matter of the third chapter. Energy generated in the deep interior of the Sun comes out as sunlight. During the passage of energy from the interior to the surface of the Sun, this energy is transported by various modes. One such mode is convection, which interacts with the magnetic fields found in the Sun to produce "blemishes" on the face of the Sun, a fact that led to much consternation in medieval Europe. The myriad of magnetic features and the changes which they undergo from time to time have kept hundreds of solar astronomers busy for hundreds of years. The fourth chapter deals with these matters. In the fifth chapter, we will come to know about the important effects wrought by these magnetic features on the Earth's environment. Solar astronomers have used an incredible variety of tools to glean all this information from the Sun. Astronomers have braved mountains and mines to extract data about the Sun. They have examined every speck of sunlight collected, have tuned into the Sun's radio emission, have fired rockets loaded with ultraviolet cameras, or sent up x-ray telescopes aboard manned and unmanned satellites. Even chemical analyses of tons of dry-cleaning fluid kept deep underground, and the use of thousands of electronic cameras looking into huge tanks of water under the ground, have now become a part of routine solar observations.

2 VITAL STATISTICS

2.1. Introduction

In this chapter, we will see how we can estimate the gross physical properties of the Sun. For one, we can clearly see that it is round in shape. This shape does not change as we view the Sun from different vantage points when the Earth performs its yearly revolution around the Sun. Thus, we can have a safe bet on the fact that the Sun is a globe or sphere. However, how large a sphere is it? In addition, how much matter does it contain? What is the average density of solar material? How hot is it? There are ways to measure these quantities without going anywhere near the Sun. However, the main measurement on which hinge all the other things is the measurement of the distance to the Sun. Many of you might now know that the Sun is 140 million kilometers away from us. However, how did astronomers measure this distance — called the astronomical unit — in the first place?

2.2. Measuring the Astronomical Unit

Kepler's third law asserts that if you multiply the cube of the size of the orbit by the square of the period, then you will get a value that is the same for all planets. Thus, the relative sizes of the planetary orbits were known right from Kepler's time. What were not known were their absolute values. Incidentally, the astronomical unit is very nearly equal to half the size of the Earth's orbit. The ellipses of Kepler are very nearly circles with very small elongations. Once we know the value of the astronomical unit, we can estimate the true

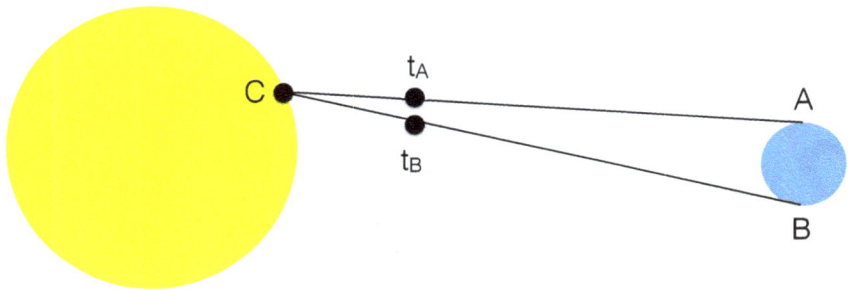

t_A is the time at which the observer at A sees the planet at the solar limb C

t_B is the time at which the observer at B sees the planet at the solar limb C

The angle ACB = $2\pi(t_A\text{-}t_B)/P$, where P is the period of the planet.

The distance CB or CA = AB/ACB = AB.P/$(2\pi(t_A\text{-}t_B))$

Figure 2.1. Triangulation is the method of measuring the distance to remote objects by measuring the two angles made by the remote object to a baseline of known length. We know that by having two base angles and one base side, we can complete the triangle and thus accurately measure the distance to the vertex. In the case of an inner planet (small black circle) orbiting the Sun (yellow circle), we can find out the times t_A and t_B when the planet is in line with the edge of the Sun, as seen from the two places A and B on the Earth (blue circle). The angle between these two planetary positions is given by the angular speed of the planet $2\pi/P$ (P is the planet's period of revolution around the Sun) multiplied by the difference in time $(t_A - t_B)$. Finally, the distance to the Sun is given by the distance AB divided by this angle. The formula for the distance is given in the figure.

size of the solar system. The famous astronomer Edmund Halley had the bright idea of using the transits of Mercury or Venus across the Sun's face to get this (Figure 2.1).

There were numerous attempts to observe the transits of Venus. These transits are also rare events, occurring in pairs more than a century apart. A modern transit observed the world over occurred on June 6, 2012 (see Figure 2.2 for the picture obtained by the GONG telescope at Mauna Loa, Hawaii, USA). Another transit was observed in the light of H-alpha at the Udaipur Solar Observatory on June 8, 2004 (Figure 2.3).

The transit of Venus was observed in 1769 from Tahiti by the expedition led by the famous Captain James Cook. He went on, after the observations, to discover the continent of Australia. The

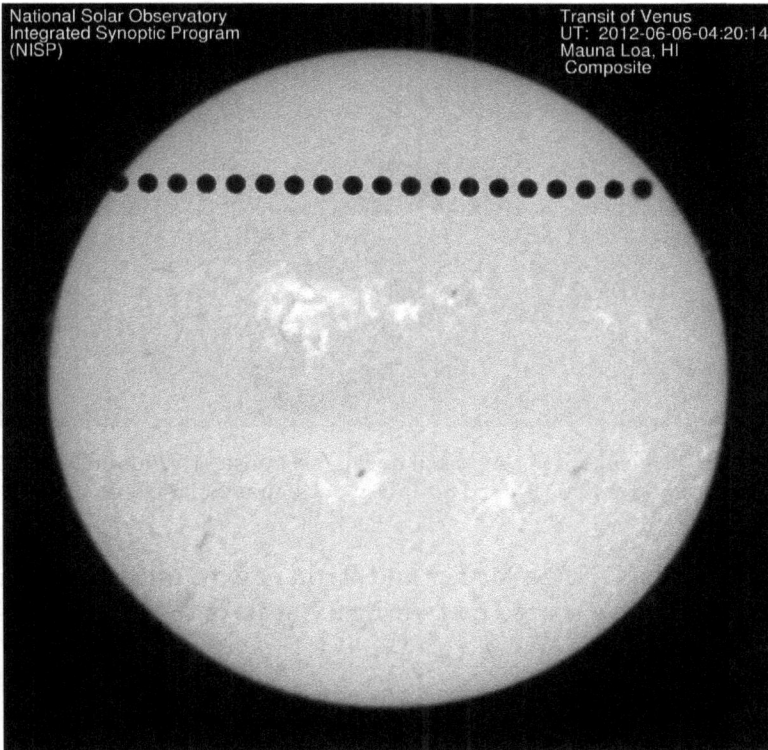

National Solar Observatory
Integrated Synoptic Program
(NISP)

Transit of Venus
UT: 2012-06-06-04:20:14
Mauna Loa, HI
Composite

Figure 2.2. Composite picture obtained by the GONG telescope on June 6, 2012, at Mauna Loa, by superposing the pictures of the transit obtained at different times. (Credit: NSO/AURA/NSF.)

accuracy of the measurement was limited by the accuracy of the chronometers that had to be carried on board the ships that went for the transit expeditions. Finally, the more accurate value was obtained by applying the method of parallax to the asteroid called Eros. More recent measurements included timing the echo of radio waves sent by giant radio telescopes and received after bouncing back from the atmosphere of Venus. Nowadays, we can also use the transmissions from space probes as they pass near planets to obtain very accurate values. The knowledge of the astronomical unit provided a baseline to measure the distances to the other nearby stars, again by the method of triangulation. From then on,

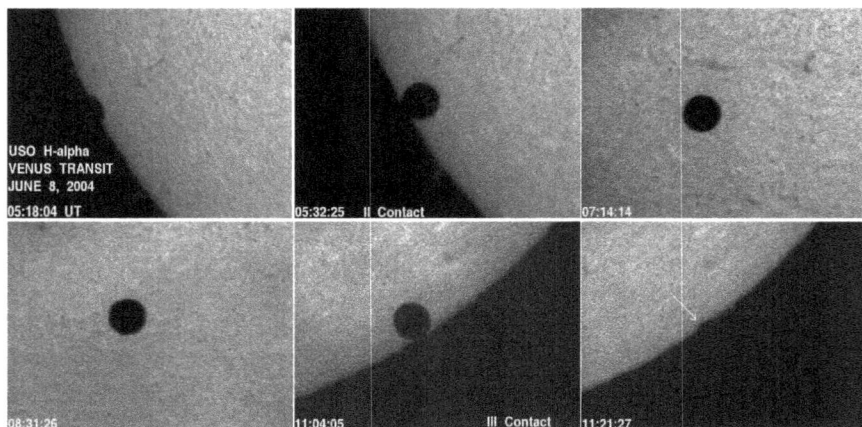

Figure 2.3.   Sequence of images taken during the transit of Venus on June 8, 2004 at the Udaipur Solar Observatory. (Courtesy of Udaipur Solar Observatory.)

the horizon has shifted farther and farther away, until modern day astronomers have started measuring the scale of the Universe itself.

2.3. Weighing the Sun

The famous astronomer Johannes Kepler put a certain order into the large amount of data collected by Tyco Brahe and arrived at his famous laws of planetary motions. Sir Isaac Newton pondered about these laws and was able to simplify things much further. He proposed that every object attracted every other object toward itself with a force that depends inversely on the square of the distance separating the objects and directly as the product of the masses of the two objects. Newton called this force gravitation. How can we use the laws of gravitation to measure the mass of an object? We can compare two masses by comparing the force exerted by the Earth's gravity on each of these masses. This is exactly the way how a common balance works.

But we want to measure the absolute masses of objects from the acceleration they produce on test bodies pulled by the force of gravity toward the massive objects. For that, we must first find out the gravitational attraction between objects whose masses are already known. From this value of the gravitational force, we can calculate the gravitational force between unit masses separated by unit distance. This particular value is called the universal gravitational constant and is represented in physics textbooks by the symbol G. Once we know the value of G, we can find the mass of any object by measuring the acceleration produced by this object on any test object.

The force of gravitation is weak and therefore difficult to measure in the laboratory. Lord Cavendish managed to measure it using a torsion balance. (Read about Cavendish's experiment in any physics textbook.) Newton showed that the regular orbits of the planets were a result of the gravitational attraction of the planets toward the Sun. The simple nature of these orbits is a direct result of the fact that the solar mass is so huge compared to that of the planets that the mutual attraction between planets barely influences their motions. If any few of the planets had possessed large masses, then the planetary orbits would not have been so systematic and regular. An extreme example is the not-so-predictable orbits of stars within a galaxy.

Kepler's third law states that the period of revolution of any planet depends only on the size of its orbit — in other words, on the separation between the Sun and the planet. It did not depend on the mass of the planet. How can we understand this? This is similar to the fact that the period of a simple pendulum does not depend on the mass of the bob but only on the length of the string. In the case of the planets, we must replace the length of the pendulum by the distance of the planet from the Sun. Then, what does take the place of the string of the simple pendulum? What invisible string pulls at the planets and keeps them from flying apart? The answer is the force of gravity exerted by the Sun on the planets.

Let us now imagine a fictitious solar system where the planets are not revolving around the Sun, but are stationary. The fact is that the planets cannot be kept in place. The pull of the Sun's gravity will make them drop toward the Sun, much in the same way as an apple would from a tree toward the ground. To keep a planet from falling into the Sun, it must have a sideways motion to make it go around the Sun. If the sideways motion has a small speed, then the planet will spiral with a smaller and smaller radius around the Sun and eventually fall into it. If the speed is large, then the planet will spiral outward with a larger and larger orbital radius until it eventually escapes from the solar system. But if the speed is just right, then the planet remains trapped in a fixed orbit around the Sun. Because the planet changes its direction of motion all the time in order to maintain a closed orbit around the Sun, we know that it is being accelerated all the time. Such an acceleration, which is always directed toward the center of the orbit, is called a centripetal ("center-seeking" in Latin) acceleration. The centripetal acceleration is provided by the gravitational attraction of the Sun on the planet. Thus, the period of the orbit is a measure of the solar gravity. So, now, we do have a measure of solar gravity at a known distance from the attracting object, viz. the Sun. Using Newton's laws of gravitation, and the value of G, we can determine the solar mass (Figure 2.4). This turns out to be a million, million, million, million, million kilograms (10^{30} kg). Now, that is quite a lot of matter indeed!

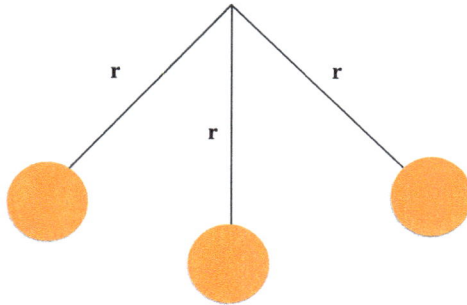

$P = 2\pi(r/g)^{1/2;}$ [Simple Pendulum]
$g = GM/r^2$ [Newton's Law of Gravitation]
$P = 2\pi(r^3/GM)^{1/2}$ [Kepler's 3rd Law]
$M = 4\pi^2 r^3/(GP^2)$

Figure 2.4.   A simple pendulum gives the relation between the length of a string and gravity. The solar system has planets held to the Sun by solar gravity. Thus, the periods of planets from Kepler's laws give solar gravity and hence solar mass.

2.4. The Temperature of the Sun

Any object, if heated to a high temperature, emits light. We see this in the case of an incandescent lamp (your ordinary out-of-fashion light bulb). Here we can heat the filament of the bulb by passing an electric current. The hot filament glows as a result. The light emitted by the lamp consists of a number of colors. However, not all the colors will be present with equal brightness. One particular color (wavelength) will be the brightest for every value of the temperature of the filament. This dominant color will range from a dull red for low temperatures to a bright yellow for hotter filaments. The temperature of the filament can be guessed from its color. Now, the dominant color of sunlight is green (see figure blackbody for the plot of energy versus wavelength of radiation). Using the same principle as in the case of the bulb's filament, we can make a pretty good estimate of the Sun's temperature to be in the region of 6000 degrees Kelvin (add 273 to the temperature in degrees Celsius to get the value in degrees Kelvin). This is the temperature of the material on the surface of the Sun (Figure 2.5). As we will see later, the temperature inside the Sun is much higher.

2.5. The Density of the Solar Material

Once we know the distance to the Sun, we can calculate the size of the Sun. A simple way is to cover a mirror with a piece of black paper which has a small hole. Now you can reflect sunlight onto a wall. A circular image of the Sun will be seen on the wall. The image will be larger if the distance of the mirror from the wall is larger. If you measure the size of the image, you will find it to be about a hundredth of the distance of the wall from the mirror. The diameter of the Sun is also a hundredth of the astronomical unit. This is because the angle made by the Sun to the pinhole is the same as the angle made by the Sun's image on the wall to the pinhole. Having estimated the diameter, you can find out the volume of the Sun. From the volume and the mass, you can estimate the average density of the solar material. This turns out to be about 300 kg per cubic meter, which is a third of the density of water in its normal

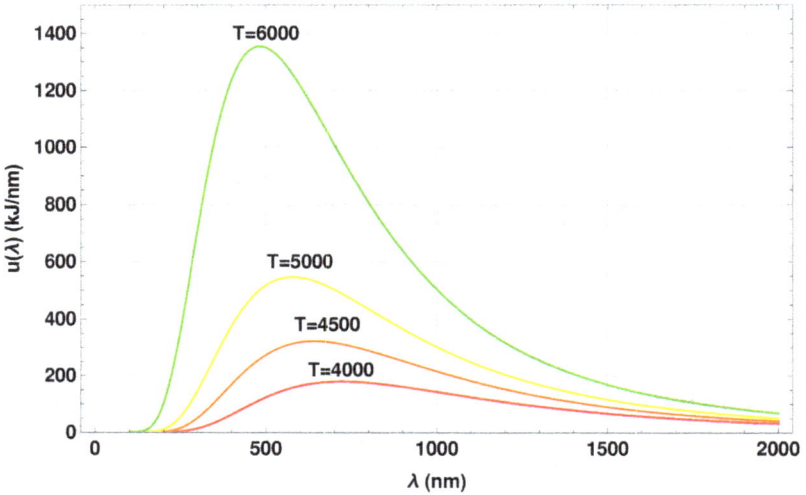

Figure 2.5.   Plot of the brightness of hot bodies as the function of wavelength. The brightness is maximum at a particular wavelength for a particular temperature. For the Sun, the peak brightness is in the green wavelength, so we can infer the temperature as 6000 degrees K. (Courtesy of Dr. Avijeet Prasad, Udaipur Solar Observatory.)

state. It is difficult to imagine the state of matter that has this density at a temperature of 6000 degrees Kelvin. Actually, this will be possible in a fourth state of matter called the plasma state. You will hear more about plasma material in the next chapter.

2.6. The Composition of the Sun

By analyzing sunlight using a device called a spectrograph, physicists were also able to find out that the most abundant element in the Sun is hydrogen, followed by helium. In fact, the element helium was first detected in the spectrum of the eclipsed Sun observed by Pogson, along with Jansen and Lockyer, in 1868 from the tobacco fields of Guntur in India, and derives its name from *helios* (the Sun in Greek). The solar spectrum shows many dark lines, known as Fraunhofer lines (after the discoverer). Each line can be identified to be caused by absorption by a particular atom (Figure 2.6).

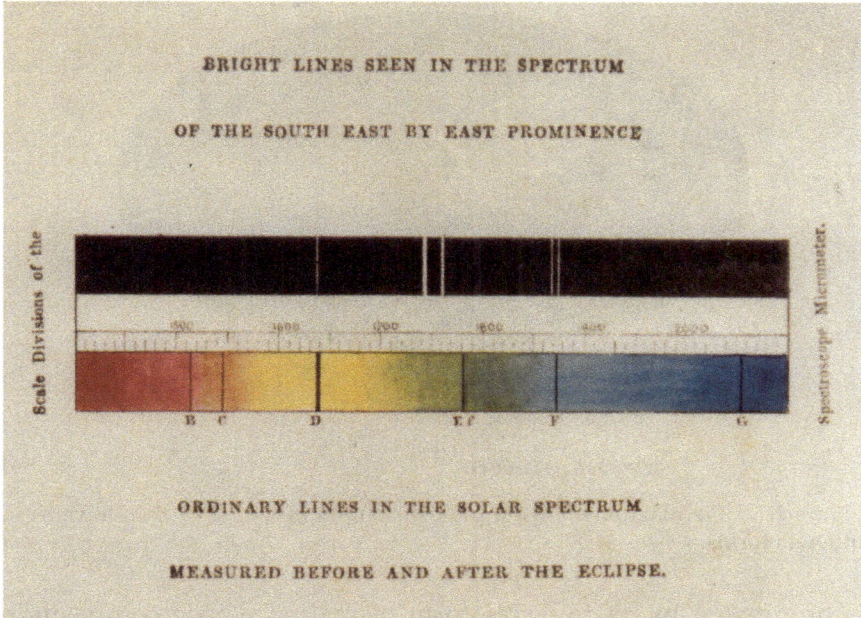

Figure 2.6.   A flash spectrum taken during the total solar eclipse in 1868, showing the then unidentified helium line in the yellow portion of the spectrum. The spectrum of the uneclipsed Sun is shown below the flash spectrum for comparison. Note that the flash spectrum has no background continuous spectrum while the normal solar spectrum has a continuous background of all colors of the rainbow superposed with the black absorption lines caused by different elements, and known as Fraunhofer lines. (Courtesy of Indian Institute of Astrophysics Archives, Bengaluru.)

It is interesting to note that the elements like carbon, iron, and silicon, which are abundant on land on the Earth, are present only in traces in the Sun. Most of the visible Universe consists of elements present in more or less the same abundance as seen in the Sun. Incidentally, Nature seems to be playing a special game with man. We have become so used to the abundance of the elements as seen on land that we tend to be surprised when we find hydrogen to be so abundant in the oceans, and much more so in the visible Universe. Now, we seem to be heading toward another big surprise, because astronomers are beginning to get evidence that most of the Universe

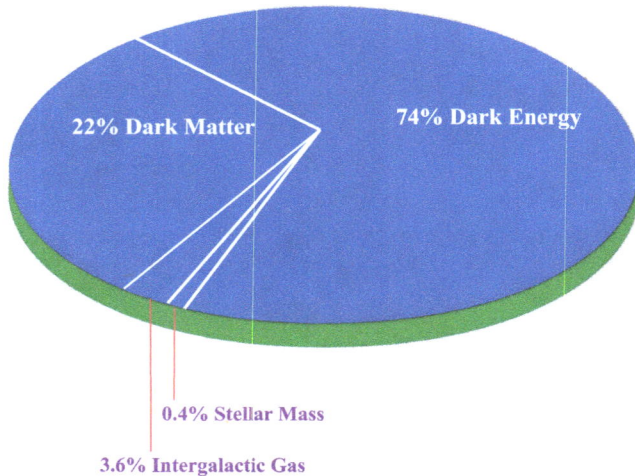

Figure 2.7.   Pie diagram showing the distribution of matter of the Universe in different forms.

is dominated by an invisible form of matter, whose composition could be very different from that of the visible Universe (Figure 2.7).

2.7. The Solar Energy Output

The most striking feature of our daytime star is its intense emission of heat and light. It will be interesting to measure the amount of energy radiated by the Sun per second. In addition, this is one of the things that we can measure without going anywhere near the Sun. First of all, imagine that the Sun radiates this energy equally in all directions. So, if we can place an imaginary energy collector shaped like a spherical shell around the Sun, then the energy radiated by the Sun per second can be measured if we collect all the energy falling on our imaginary collector. If we measure the energy crossing unit area of this shell in one second, and then multiply this with the total area of the collector, then we will get a very good estimate of the total energy emitted by the Sun in one second. We know that the total area of the collector's surface is $4\pi r^2$. If the collector were to have the size of the Earth's orbit around the Sun, then r would equal the astronomical unit. So, if E is the solar energy falling on 1 square meter per second of an object kept on the Earth, then the total solar output is given by $L = 4\pi r^2 E$.

What would be a simple way of collecting the solar energy E? The simplest way that I can think of is to coat a thick copper plate with black soot and place it in the sunlight. Because of the soot, the plate will be able to absorb almost all of the sunlight falling on it. The temperature of the plate will first rise, because the absorption of energy will make it hot. As the temperature rises, the plate will start radiating away some energy and will try to cool off. Soon, the rate at which it tries to cool off by radiating the energy will equal the rate at which it absorbs the energy. When this happens, the temperature will not rise, but will reach a steady value. If you take some care at the beginning and apply the soot very carefully, then you will find that the plate will reach a temperature of about 50 degrees Celsius in summer. Now, completely absorbing bodies are called blackbodies by physicists, and there is a definite law relating the temperature emitted by a blackbody and the amount of energy radiated from 1 square meter of its surface. But because the temperature has reached a steady value, we know that the amount of energy radiated by the plate per second from 1 square meter of its surface is exactly equal to the amount of energy absorbed onto 1 square meter of the plate per second. Because the soot absorbs all the energy falling on it, we can also take this absorbed energy to be exactly equal to the energy falling on 1 square meter of the plate per second. This is just the data we need. To calculate the energy radiated by 1 square meter of a blackbody per second, we have to first add 273 to the temperature in degrees Celsius, then multiply the result with itself to get the square of the temperature, and then multiply the square with itself to get the fourth power of the temperature. We must finally multiply the result by the value of a constant called the Stefan–Boltzmann constant, to get the energy in watts per square meter. If you do all the calculations correctly, you will see that the energy output from the Sun is about a hundred million, billion, billion watts! Considering the fact that this energy must have been liberated from less than a thousand billion, billion, billion, kilograms of matter for 4.5 billion years (which is the age of the Sun), we arrive at the straightforward conclusion that the calorific value of the solar material must be more than ten million calories per kilogram.

Physical property	Value
Mass	$2 \cdot 10^{30}$ kg
Radius	$7 \cdot 10^8$ m
Mean density	1400 kg m^{-3}
Surface gravity	100 m s^{-2}

We now come to the end of this chapter. We have seen that simple physical considerations lead to definite information about the vital statistics of the Sun. In addition, what we learn is somewhat disquieting. How is this hunk of plasma held together? How is it able to generate such enormous quantities of energy? Can this state of affairs go on forever? These questions do arise when we confront the vital facts about the Sun. We will discuss these questions in the next chapter.

3 THE ANATOMY OF THE SUN

3.1. Introduction

We have seen in the previous chapter how much energy is radiated by the Sun. We also know how difficult it is to produce even a small fraction of this energy on the Earth. One of the greatest problems associated with the increasing sophistication of our way of life and the increasing demands created by the soaring population is the production of sufficient amounts of energy. It is not surprising, therefore, that many brilliant minds pondered about the source of the Sun's energy and physicists now have a good idea of the mechanism which produces this energy. There is a story attributed to Hans Bethe (who had identified the source of energy of the Sun and the stars) about an evening spent looking at the night sky with his girlfriend. After staring at the stars for a long time, he turned toward her. She expected some romantic remarks — but Bethe confided that he was the only person in the world at that time who knew what made the stars shine!

Before we can go into the details discovered by Bethe half a century ago, we need to find out what the Sun is made up of, and how it is held together. These facts have an important bearing on the way the Sun produces its energy. By analogy, it is the same need for physicians to study the anatomy of the human body, in order to understand how we live and move about.

3.2. The States of Matter

When holding a stone or a piece of wood in your hand, have you not wondered what it consists of, and how it retains its shape? Likewise, why do liquids need a container to keep them in place, and why does a gas always expand to fill its container? The answers to these questions will help us understand the different states of matter. After doing this, we can go ahead and try to deal with the stuff that the Sun is made up of.

All matter is made up of fundamental building blocks, called molecules. In dealing with matter, we normally like to look separately at the physical properties and the chemical properties. The hardness of a substance, the way it responds to heat or electricity or magnetism — these are some examples of the physical properties. The fact that a substance is in a solid, liquid, or gaseous state depends on its physical properties. On the other hand, the purity of a substance is something related to its chemical properties. A sample of water is pure if it consists of water molecules and nothing else. The same number of water molecules can group together in a solid form or solid state, namely ice. Water in its normal state is a liquid. Steam is a collection of water molecules in a gaseous state.

The water molecule itself is made up of two hydrogen atoms and one oxygen atom. Atoms are the fundamental building blocks of special substances called elements. In an element, you will find only atoms of one kind. Different chemical elements combine in different proportions to form all the substances found in nature. The atoms themselves are made up of a central nucleus (with a positive electric charge) surrounded by a swarm of electrons (which are negatively charged). These particles are made up of other things called quarks. It is simple to ask whether the quarks are made up of smaller things, but cutting up particles into smaller and smaller ones requires particle beams of higher and higher energy. These, in turn, require more and more expensive experiments, thus making progress in this area somewhat sluggish. The Large Hadron Collider is the most recent experiment which has created much uproar in the media — because of reports that this experiment might create black holes which could destroy the Earth! This reminds me of the uproar created in medieval times when Galileo discovered sunspots. Humans do not change, do they?

Let us get back to the states of matter. The reason why the molecules of a substance stick together is that they are all attracted to each other by the intermolecular force (IMF). A solid is compact and retains its shape because the IMF is strong in a solid. A liquid cannot retain its shape because the IMF is weak. However, there is still some force. Therefore, the liquid maintains its size, but not its shape by easily adapting to the shape of its solid container. In a gas, the IMF is practically zero, allowing the molecules to move about freely. That is why a gas always fills the entire volume of its solid container in a uniform way.

Coming back to the water molecules, you may well ask why the molecules should have a stronger attraction in a solid which becomes weaker in a liquid or a gas. The answer is that the attraction decreases when the molecules are farther apart. This answer is not so complete, however, because we face a new question as to why the molecules in a solid are more closely packed than in a liquid. The answer to this new question lies in the fact that the molecules are not motionless, but keep moving about. You know well that the energy of a moving object is more than that of a stationary one. Anybody hit by a cricket ball hurled by a fast bowler will be painfully aware of this fact! Now, this energy is the kinetic energy. The kinetic energy of a molecule is greatest when it is part of a gaseous substance and it is least when the molecule is in a solid. When a molecule has less kinetic energy, it moves about less violently. Because of this, the distance it moves from any particular position is also smaller. When you put a bunch of such lethargic molecules together, they will maintain a smaller separation on the average, and therefore attract each other more strongly. They thus form a solid.

When you heat a solid substance, each molecule absorbs a portion of this heat energy and starts to move about more violently. The molecules now have sufficient kinetic energy to travel a little farther apart. Their average separation increases. What we will see on the outside is an expansion of the body. So, we will conclude that a solid expands on heating. Heating the body further will eventually melt it into a liquid. At this point, the kinetic energy of the molecules will have increased a lot, pushing them so far apart that only a weak IMF prevails. Further addition of heat makes the molecules so violent that they break away from all intermolecular attraction, and become freely moving molecules. They now exist in the gaseous state.

Suppose you are able to tamper with the internal structure of a molecule and succeed in liberating one or more electrons from its clutches. By removing a negatively charged electron from a neutral molecule, you will be left with a positively charged thing called a positive ion. Positive ions and negative electrons attract each other with electrical forces. When they are close to each other, each positive ion will attract only its daughter electron. However, if you are able to give some extra kinetic energy to the electron, it will move somewhat far away from its parent ion. It will now find that it is weakly attracted to many other ions in its neighborhood. In the same way, the parent ion will now attract many other electrons apart from its daughter electron. This collective attraction between many ions and many electrons produces interesting electrical and magnetic phenomena in the matter containing such weakly attracting electrons and ions. This state of matter is the plasma state. The weak attraction lasts only out to a short distance from each ion or electron. This distance is called the Debye length. Beyond this distance, the attraction is practically zero and the electrons and ions are as free as the molecules of a normal gas. A simple example of plasma is a candle flame. Another example is the solar plasma.

But hold on! Solids, liquids, and gases are states of matter of increasing kinetic energy for their constituent molecules. The kinetic energy of the constituent ions and electrons in the plasma state would have to be larger than that of the molecules of a normal gas. However, gas molecules are themselves so energetic as to be free from intermolecular attraction. They would fly apart unless trapped within a container. Therefore, a plasma will also need a container. Although a candle flame looks so compact, and contained, it actually consists of a continuous flow of hot gas away from the flame. As the gas emits light, it loses energy and therefore cools. When the molecules move a little away, they will cool down so much that they no longer emit light and become nonluminous. Thus, the flame is observed only out to a short distance, and so looks compact, although it is not really restrained. Similarly, one could argue that the solar plasma is not as compact as it seems, but is moving out like the flame molecules. Such a migration of molecules would quickly exhaust the entire Sun within a time that is short compared to the observed age of the Sun. So, in the case of the solar plasma, unlike that of a candle flame, the plasma is really confined. So, what keeps the Sun together?

3.3. Keeping the Sun Together

If the Sun is in the plasma state, then the positive ions and electrons are practically free of each other when they move apart beyond the Debye length. From the temperature and the mean density of solar material, we can calculate the Debye length. However, the Sun is much bigger than this length. Thus, actually, the solar plasma consists of practically free particles. So the question, as asked earlier, is: What keeps these particles bound together?

The answer is gravitation. The same force of gravity that holds the planets of the solar system together also holds the otherwise free particles of the solar plasma together. You will then ask why we cannot form minisuns in the laboratory by keeping some particles bound together in the plasma state. What is so special about the Sun which disallows the plasma particles to diffuse away? After all, gravitational attraction between particles can act anywhere in the Universe — even on the Earth. Why does gas behave as it does, by always expanding to fill its container? Well, if you make a quick calculation, you will see that the force of gravity between two gas molecules is very weak. You need a sufficiently large mass of material kept together in the first place, to attract other molecules with sufficient force. This is what happens to the molecules of the Earth's atmosphere which remain stuck to the Earth without diffusing away. On the other hand, the Moon's gravity is not strong enough to retain a dense atmosphere, and so the Moon does not have a dense atmosphere. Therefore, to keep the solar material together, we need a large mass in the first place. How did this seed mass originate? That is the story of star birth, which I will not tell in this book.

Let us now probe a little more into the self-gravity of the Sun. In a solid, we expect the IMF acting on a molecule inside the object to be almost zero because it is pulled equally in all directions. On the other hand, a molecule on the surface of the object will have more molecules pulling it inward, and practically no molecules pulling it in the outward direction. In spite of this apparent imbalance of the IMF, the object does not collapse inward, but remains rigid. How? The answer is that the IMF is a force which has a short range. Thus, only a few nearest neighbors of any particular molecule will feel the IMF. So, no matter where the molecule is, whether on the surface or in the interior of the object, this molecule will be attracted only by a few neighboring molecules. There is no real imbalance in the forces on the surface of the object. The object is able to retain its shape, and is not required to collapse inward. In addition, since the compactness is the same throughout the object, the density of any sample cut from this object will be more or less the same.

The situation is completely different for a self-gravitating material. The gravitational force is a long-range force. A plasma particle on the surface of the Sun is attracted by a particle in the center of the Sun by a force which is only a fourth of the force exerted by another particle halfway in between. Considering the fact that there are such a large number of interior particles compared to the number of surface particles, we should imagine the inward pull on the surface particles to be overwhelmingly strong. If there is nothing to prevent this pull from dominating, the Sun would collapse into nothing! This strange prediction on the ultimate fate of stars could not be digested by the famous astronomer Sir Arthur Eddington, when a young Indian astrophysicist first proposed it in 1930. This young man — none other than Subrahmanyan Chandrasekhar — was awarded the Nobel Prize much later, in 1983, for realizing this astounding fact. The final product of such an unhindered collapse of stellar material is a black hole. The story of black holes is very fascinating, but I cannot fit that into this book. However, the moral of this piece of history is that nature respects no person, however eminent he or she might be. Science teaches a certain amount of objective modesty to the true student.

The Sun has existed for at least 4500 million years, as evident from the age of the Earth. So, clearly, something is preventing its collapse into a black hole. This something is nothing other than the thermal energy of the solar plasma particles that makes these move about in a random way. The steady bombardment of these particles produces a push in all directions, much like the intense battering of raindrops which you will feel during a downpour. This push per unit area is the pressure. The pressure is also equal to the thermal energy per unit volume of the material. Therefore, we want the pressure force to balance the gravitational attraction and prevent the Sun from caving in (Figure 3.1a). Since the gravitational force acts predominantly inward, we need to make the pressure force act outward. The only way of arranging this is to keep the pressure very large inside the Sun, and then let the pressure become smaller in the outer layers. The large pressure in the interior of the Sun heats up the gas in a way similar to the heating-up of the compressed air in a bicycle pump (Figure 3.1b). But energy is also escaping from the Sun in the form of sunlight. If nothing replaces this leakage of energy, then the Sun will lose all the energy gained by compression and stop radiating in 10 million years (this time scale is called the Kelvin cooling time). This time is much shorter than the 10-billion-year lifetime of the Sun. So, something must be continuously heating the central portion of the Sun. Calculations show that the temperature of the solar interior is several millions of degrees Kelvin. What is the source of energy that can heat the plasma to such a high temperature? Our understanding of how the Sun is kept together is not complete until we crack this problem. (To get a "sneak preview", see Figure 3.1c.)

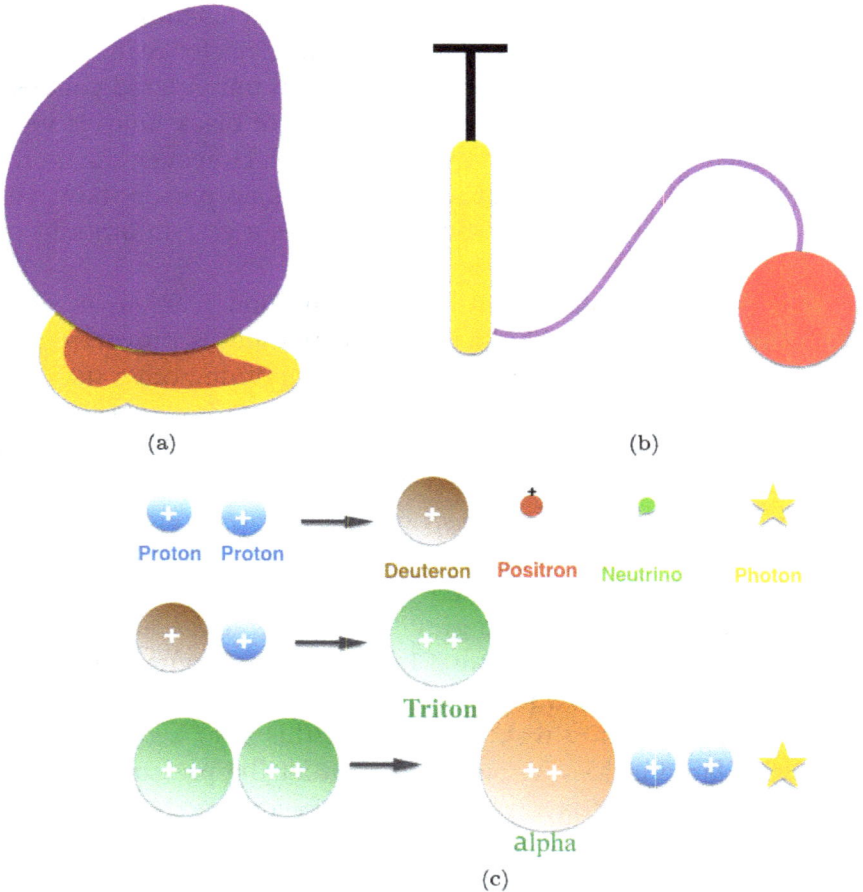

Figure 3.1.   (a)The central portions of the Sun are crushed by the weight of the overlying material (atlas.html). (b)A crushed plasma will get heated up, as we see when we pump air into a football. (c)This increases the temperature of the central portions. When the temperature becomes large enough, fusion reactions are started, converting hydrogen (proton) into helium (alpha particle).

3.4. Energy Production in the Solar Interior

Energy can be produced, in general, by various methods. On the Earth, we can do it by burning wood, coal, or petroleum products. We can also create electrical energy using the kinetic energy of moving water, as in hydroelectric power generators. In all such processes, the energy produced must be greater than the energy spent initially to start the process. Only then can such a process be a net producer of energy. Take the case of burning wood as a fuel. To start the burning, we must first light a fire. The energy required to start a small fire with a matchstick is small compared to the energy produced by the burning wood. When all the wood is converted into ash, the fire stops.

In the Sun, the energy requirement is very large. As we have seen earlier, the calorific value is in the range of 10 million calories per kilogram of material. Such prolific energy production is possible only in the case of the hydrogen bomb. It is no coincidence that the Sun produces its energy in the same way as in the hydrogen bomb. I will now try to explain this process.

In the deep interior of the Sun, the temperature of the hydrogen plasma is about 10 million degrees Kelvin. The protons move about with large speeds at this temperature. A majority of these protons have an average speed of 300 kilometers per second. Even this high speed is not enough for the proton to crash through the barrier created by the repulsive electrical force of another proton (both are positively charged). Some of the protons at this temperature will have speeds that are far in excess of the average speed. This is true for all things that are statistical in nature. For example, if you take a class of a hundred students, you will find that the vast majority of them, are average students. However, a very small number will be exceptionally bright students and will be competing for the first few ranks. In the same way, the few exceptionally energetic protons in a sample of the hydrogen plasma will be able to penetrate the repulsive force between the positive charges. There is another property of very small particles, called the quantum property. According to this, the small particles behave like waves and have a finite probability for penetrating the so-called Coulomb barrier of the repulsive force in almost a ghostly fashion. The quantum behavior of the very small particles is indeed bizarre and is very nicely illustrated in the book *Mr. Tompkins in Wonderland* by George Gamow, who proved that protons can approach quite close to each other in the interior of stars with a finite probability.

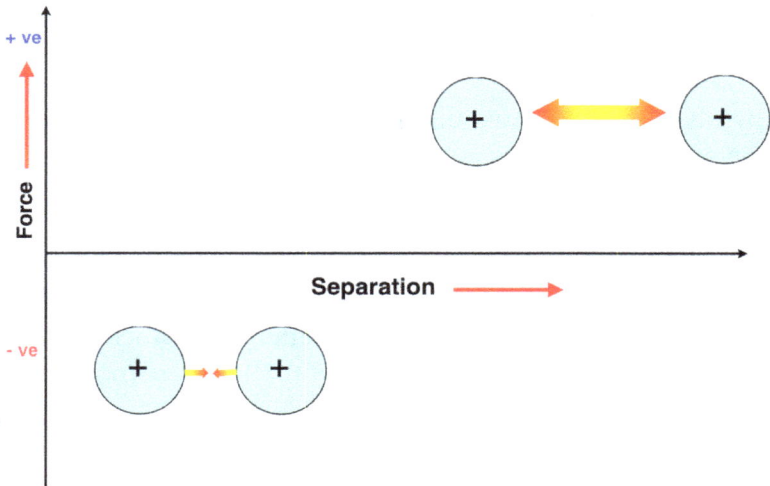

Figure 3.2.   Plot showing how the repulsive Coulomb forces between two +ve charged protons change into the attractive nuclear forces between two nucleons when the separation between the particles becomes very small, in the range of the size of a nucleus.

When such events occur, the two protons will come very close together. When their separation becomes as small as the size of the proton itself, they come under the influence of very-short-range forces that are attractive, instead of repulsive (Figure 3.2). This force, called the "strong" force, was first postulated by Yukawa to explain the binding of nuclear particles within the nucleus. The mutual "strong" attraction is not felt outside the very short range of this force.

Therefore, our two brave protons come under the spell of the strong force and fuse into a single nucleus called a deuteron. A deuteron is the nucleus of deuterium, commonly known as "heavy hydrogen". It consists of a neutron and a proton. During the conversion of two protons into a deuteron, a positively charged particle called a positron is emitted. Positrons are antiparticles. When a positron meets an electron, an annihilation of the two particles takes place. This process produces a very-high-energy photon. This kind of radiation, called gamma radiation, is used in hospitals for the treatment of cancer.

A deuteron, once produced, can also crash into a proton to produce an abnormal nucleus of helium. This nucleus contains two protons and one neutron, whereas a normal helium nucleus has two protons and two neutrons. Two abnormal nuclei of helium can also crash into each other to form one normal nucleus after releasing two protons. So, what do we have here? To produce one abnormal nucleus of helium, we require three protons collected in two separate collisions. Thus, we require six protons to produce two abnormal helium nuclei. In the final reaction, we get back two of these six protons after forming the normal helium nucleus. If we do our arithmetic right, we will see that four protons have collected to form one normal helium nucleus, also called an alpha particle. This is a fusion chain reaction.

The crashing of proton into proton produces gamma radiation along with the deuteron. In the same way, the crashing of deuteron into proton produces a gamma ray photon. The energy of these photons is larger than the kinetic energy of the protons that took part in the reaction. Thus, there is a net gain in the energy produced in this process. Such reactions that produce a net gain of energy are called exothermic reactions. How can this gain in energy be possible for the fusion chain reaction? This is possible because crashing protons kindle the hidden energy present in the "strong" forces, much in the same way as a lit matchstick can kindle a huge wood fire. However, this gain in energy must be at the expense of something. It turns out that the mass of a single alpha particle is smaller than the mass of four protons put together (Figure 3.3). So here is a case of mass being converted into energy, an example of the celebrated $E = mc^2$ principle, which was expounded by Einstein. The reactions will stop when all the available high speed protons in the hydrogen plasma convert into alpha particles.

While producing the energy, we must take care that it does not collect at some place. If a large amount of energy collects in a small region, then an explosion will result. This is exactly what happens in a firecracker or a bomb. But we are pretty sure that the Sun has not exploded away! So clearly, it must be throwing out the energy produced by nuclear fusion from the interior to the surface in a

Figure 3.3.   Four protons are more massive than an alpha particle.

very efficient way. How this transport of energy proceeds will be described in the next section.

But before I end this section, I must mention an interesting thing that happens during the first reaction, viz. that of two protons combining into a deuteron. If we do a headcount, we will see two participants in the reaction: We have the proton and neutron sitting inside the deuteron that accounts for two normal particles. However, we also have an extra antiparticle, viz. the positron. To conserve the number of normal particles on either side of the reaction, we need another normal particle to balance the positron. This particle is the neutrino. The neutrino interacts with other particles through a special kind of force, called the "weak" force. This is very weak compared to the "strong" force and therefore the neutrino is very rarely captured by the other particles. So, the neutrinos, which are produced in the nuclear reactions occurring deep in the interior of the Sun, escape without much of a problem, into outer space. The nuclear reactions involving neutrinos are called weak interactions and have a very low probability — and thus a very long reaction time, in the range of 10 billion years! It is precisely this reaction that slows down the burning of the hydrogen fuel in the interior of the Sun and Sun-like stars and accounts for the long 10-billion-year life of the stars.

Anyone clever enough to detect these neutrinos would get a direct proof of the theory that Hans Bethe proposed about nuclear reactions within stars. This has actually been done in recent times, and we will have more opportunity to learn about this in the last chapter of this book. The energy carried away by the neutrinos is smaller than the energy radiated in the form of photons by the Sun. However, for some stars this energy loss by neutrinos becomes very important. The sudden loss of energy makes the cores of these stars collapse very violently. The reflected shock wave produced by the collapse can throw out the outer layers of these stars in a dramatic, explosive way. These explosions are known as supernovae. The Crab Nebula (Figure 3.4) is an example of the debris left

Figure 3.4. Picture of the supernova remnant called the Crab Nebula. Inside this nebula resides the solid remnant of the explosion, a spinning magnetized neutron star that is known as the Crab Pulsar. (*Source*: http://wallpapered.net/tag/crab-nebula/)

behind after a supernova explosion, observed by ancient Chinese astronomers in 1054 A.D.

The supernova explosion that occurred in 1987 in a neighboring galaxy caused a lot of excitement for astronomers, because they had the first chance to study a supernova explosion at close quarters (but from a safe-enough distance!) using modern telescopes and instruments. What is more, neutrino detectors were able to detect the sudden, impulsive arrival of the neutrinos from the supernova, which conclusively proved the connection between neutrino loss and supernova explosions.

3.5. Energy Transport in the Sun

As mentioned in the previous section, we cannot let energy accumulate at one place, because that will lead to an explosion. Since the Sun is not exploding, it means that the energy produced in the deep interior of the Sun must be continuously flowing away from the Sun.

In the innermost region of the Sun, a sphere with a third of the Sun's radius, the temperature is hot enough to sustain thermonuclear fusion reactions. Outside this sphere, the temperature is lower and no energy is produced. In this region, the gamma rays released from the core of the Sun are captured by atoms, which reradiate the gamma rays. Not all the gamma radiation is captured in one layer, but eventually, after passing through many such layers, all the gamma rays are caught. The average distance for which a gamma ray photon can travel in the plasma before it is captured is called its mean free path. The mean free path depends on the ability of the plasma to capture photons. This ability is called opacity, a word derived from "opaque." A substance with large opacity will be opaque to the radiation and will not allow it to pass through. The mean free path of the photon will be small if the opacity is large. The opacity depends on the number of absorbing atoms as well as on the size of the photon, technically known as the wavelength of radiation. For example, bodies opaque to normal, visible light can allow x-rays or gamma rays to pass through. The mean free path of the photons will be smaller in the deeper layers of the

Sun, compared to the outer layers, because the absorbing atoms are packed more densely in the deeper layers.

A photon, once absorbed, is re-emitted at a wavelength that is equal to or longer than that of the absorbed photons. Those emitted with equal wavelengths are absorbed again. Thus, a result of many such absorption encounters is a gradual buildup in the number of photons of longer wavelengths. Gamma ray photons are converted into x-ray photons, then into ultraviolet photons, and finally into the photons of visible light. It so happens that the mean free path of the photons becomes larger than the thickness of that layer where visible photons are abundant. These photons have a greater chance of escaping to outer layers because of their large mean free paths. Since the outer layers are progressively less dense, the mean free path in the outer layers is even longer. This starts a steady leak of photons away from the Sun. The sunlight, now free to travel without being absorbed, permeates into outer space, falling on the Earth, the planets and any other exo-planet orbiting a distant star. An intelligent observer on this exo-planet will see our Sun like a twinkling star and might wonder, like Hans Bethe, how it produces the light. The layer on the Sun from where sunlight leaks through is called the photosphere. We thus see that the photons are able to carry away the energy from the core by a kind of drunkard's walk, hitting one atom, then another, and so on. This process is called radiative diffusion. It takes about one million years, on average, for the energy to diffuse out from the Sun.

Even this slow rate of diffusion is not equally efficient at all the layers. The temperature of a layer roughly 200,000 kilometers below the solar surface is about a million degrees Kelvin, and lower than this for the outer layers. At this temperature of a million degrees, the mean energy of a plasma particle is greater than the binding energy of a proton with an electron to form a neutral hydrogen atom. This energy is called the ionization potential. For layers with smaller temperatures, the average energy of the plasma particles drops below the ionization potential. Now, the particles start combining to form neutral hydrogen atoms. A photon hitting a neutral atom is strongly absorbed by the atom. Instead of reradiating another photon, all that happens is that the energy of the photon is completely spent in ionizing the atom. At this stage, the free diffusion of photons is hampered, and the danger of energy accumulating becomes imminent.

This stagnation of energy leads to a local increase in temperature. This produces an interesting effect. The same effect is seen when you heat a pan of water. The layers of water closer to the fire get heated first. These heated layers expand and become lighter than the overlying layers. Therefore, they rise, and are replaced by cooler, heavier layers from the top. In this way, a kind of rolling motion of water sets in, growing into a bubbling motion. This constant churning of the fluid, replacing the lower, heated layers with cooler layers from above, is called convection and is an efficient way of transporting heat from the bottom to the top. The Sun, too, undergoes such convection, where the layers heated by energy stagnation at a depth of 200,000 kilometers are made to rise to the surface. The time taken for one such turning-over of the solar plasma is roughly one month. The region where such a mode of energy transport takes place is called the solar convection zone.

Apart from transporting energy, the convection zone produces many sound waves, much in the same way as the water in a heated pan produces a splashing sound while boiling. These sound waves cannot be detected by the human ear (if a human could approach near enough!), because they have a frequency of a few thousandths of a cycle per second, whereas human ears respond to a frequency range from a few hundred to a few thousand cycles per second. The solar sound waves play a vital role in the heating of the outer layers of the solar atmosphere. The sound waves generated in the convection zone of the Sun also excite oscillations of the entire Sun. These global oscillations of the Sun can be used as a probe of the solar interior, much in the same way as the clarity of the tone of a bell can be used to detect whether the material of the bell has any flaw. Of course, the sound waves cannot cross the near vacuum of interplanetary space. They are detected by measuring the up and down motions on the solar surface caused by the sound waves using the shift of spectral lines created by the Doppler effect in light waves.

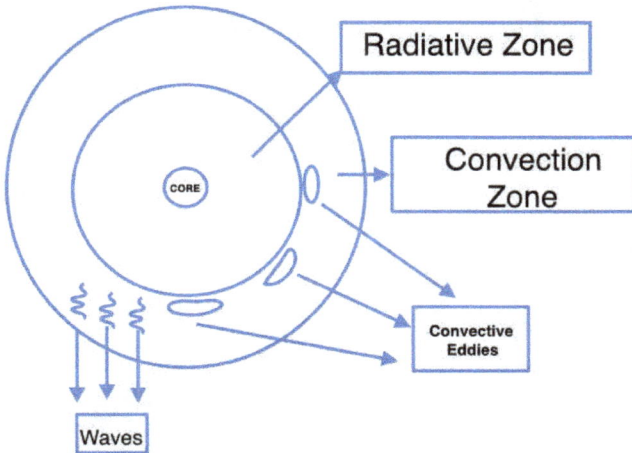

Figure 3.5. The innermost 30% is the energy-generating core, surrounded by the radiative zone up to 0.7 of the solar radius. Convection with eddies or bubbles of plasma sets in at this radius when the radiation is prevented from flowing out. These eddies generate sound waves some of which are trapped as standing waves whose frequency distribution can be used to estimate the interior solar properties with great accuracy. The energy again escapes from the surface of the Sun in the form of photons. Hence, the surface layer is called the photosphere.

The anatomy of the Sun is summarized in Figure 3.5. The innermost 30% of the Sun is the energy-producing core, surrounded by the radiative zone up to 70% of the solar radius. Convection with eddies or bubbles of plasma sets in at this radius when the radiation is prevented from flowing out. These eddies generate sound waves, some of which are trapped as standing waves whose frequency distribution can be used to estimate the interior solar properties with great accuracy. The energy again escapes from the surface of the Sun in the form of photons. Hence, the surface layer is called the photosphere.

4 BLEMISHES ON THE SUN

4.1. Introduction

Whatever you have learnt so far about the Sun was based purely on physical principles. The story so far could have applied to any one of the millions of stars that populate the night sky. The measurement of flux, angular diameter, and dominant color of the star's radiation is possible for many stars. Thus, the luminosity and temperature of such stars can be calculated. Based on these values, we can find out the way these stars are held together, the mode of energy generation, and the mode of energy transport. Supposing that the Sun is very far away, and is visible as a tiny star, we could have ended our studies with the previous chapter. The photons leaking from the photosphere have too large a mean free path to have any interaction with the plasma particles of the overlying atmosphere. These layers would therefore be not heated by the outward flow of photons. They would be cooler and cooler at greater heights from the solar surface. The properties of such an atmosphere could have been easily predicted. In fact, there is a theorem in astrophysics called the Vogt–Russell theorem, which states very confidently that the structure of a star can be completely determined once we know its mass and chemical composition. Alas, this confidence is not fully justified. The reason is that the atmospheres of two stars with the same mass, luminosity, and chemical composition are sometimes seen to be very different from each other. What causes such differences? One source of the differences could be the differences in the magnetic field strength of the star. A study of the Sun's magnetic field and its relationship with the features seen on the Sun's surface will eventually help us to understand the similar processes occurring on other stars.

4.2. Sunspots and Their 11-Year Cycle

The first inkling of the surprises in store for astronomers came with the invention of the telescope. Galileo trained a telescope at the Sun, and found that the Sun's face was not pure white, but had several dark spots. These blemishes on the face of a heavenly body caused a lot of confusion for the religious leaders of that time, who always imagined that heavenly bodies were free of defects. However, I wonder why that much fuss was not made about the craters on the Moon, which is easily visible to the unaided eye? A white light full disk picture showing sunspots (Figure 4.1) and an enlarged view of many sunspot groups (Figure 4.2) are shown below. The dark central portion of each spot is called the umbra, while the lighter peripheral region is called the penumbra. The penumbra is made up of a large number of elongated fine structures called penumbral filaments. These filaments are caused by the bubbling convection in the presence of an inclined magnetic field. Notice the regular pattern of granular convection or granulation in the region surrounding the sunspot where there are no strong large-scale magnetic fields.

The dark sunspots remained a curiosity for many years. Careful recording of the sunspots' positions, day after day, for many years by many scientists, like Carrington, Maunder, Wolf, and Sporer, revealed a curious waxing and waning of the number of sunspots with a period of about 11 years (Figure 4.3). I once again refer to the excellent book *The Sun Kings* by Stuart Clark, which narrates the fascinating story of solar activity and its influence on the Earth's atmosphere. Many questions arise when we read about the spots and their periodic appearance. What makes these spots increase and decrease in number? Why are the spots dark? Why do they sometimes not appear at all, as in 1650–1750, a period called the Maunder minimum?

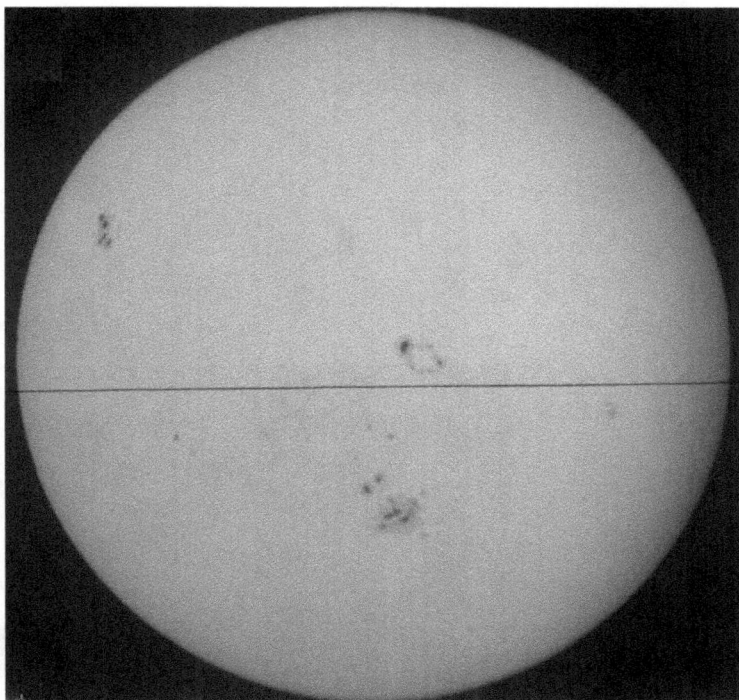

Figure 4.1. A full disk image of the Sun taken in white light (broad band visible light). The several dark blemishes seen on its surface are actually groups of sunspots. (Courtesy of Indian Institute of Astrophysics.)

Sometimes, scientific understanding proceeds in strange ways. Zeeman discovered, in 1898, that spectral lines formed in magnetic fields split into many components, and he was awarded the Nobel Prize for this discovery in 1901.

Across the Atlantic, George Ellery Hale, of Mount Wilson Observatories in the U.S., had been recording pictures of sunspots in a spectral line of hydrogen at 656.3 nanometers. In fact, he (and, independently, Jansen of France) had developed a special instrument called the spectroheliograph for doing this. Hale noticed whirlpool-like structures emanating from the sunspots (see Figure 4.4). He was reminded of the distribution of iron filings around a magnet, and wondered whether sunspots had magnetic fields. The Zeeman effect now gave Hale a way to detect magnetic

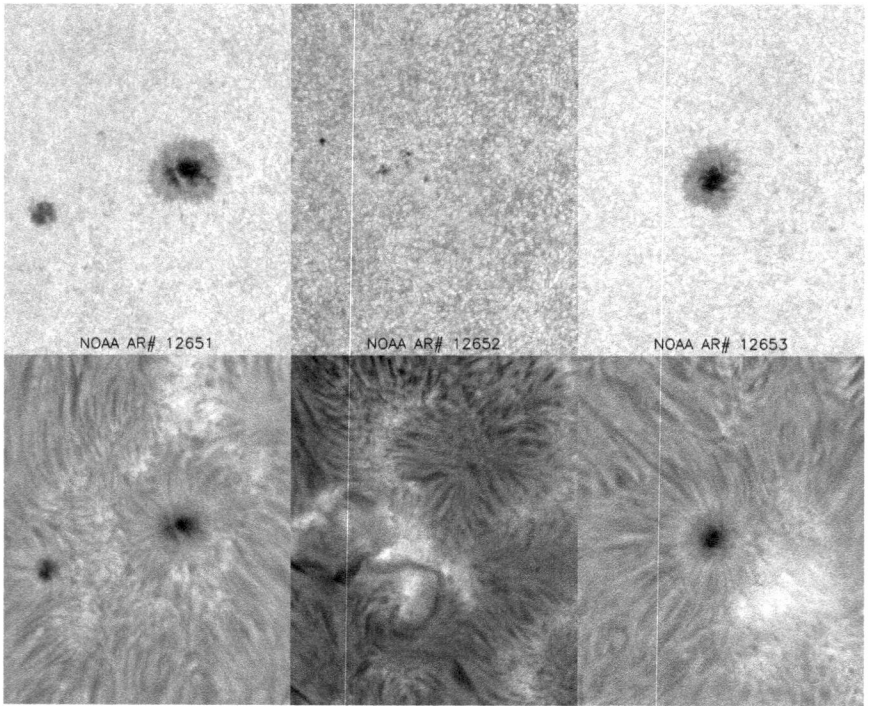

Figure 4.2.   Pictures of different sunspot groups obtained at the newly installed Multi Application Solar Telescope (MAST) of the Udaipur Solar Observatory. *Upper row*: Different kinds of sunspots seen in the G-band light which comes from the photosphere. *Lower row*: View of the corresponding regions seen in the H-alpha line which comes from the chromosphere. (Courtesy of Udaipur Solar Observatory/Physical Research Laboratory.)

fields in sunspots. He obtained the spectrum of a sunspot in a magnetically sensitive Fe line and indeed found a splitting of spectral lines in 1908 (see Figure 4.5). He could precisely measure magnetic field strength in the sunspots and found it to be 0.1 tesla, which is 10,000 times stronger than the Earth's magnetic field.

John Evershed, from the Kodaikanal Observatory in India, wanted to see whether the sunspot's magnetic field was being produced by circular currents created by a cyclonic motion of gases as inferred by the whirlpool structure of sunspot whorls. He went about measuring the motions of the gases using a spectrograph,

DAILY SUNSPOT AREA AVERAGED OVER INDIVIDUAL SOLAR ROTATIONS

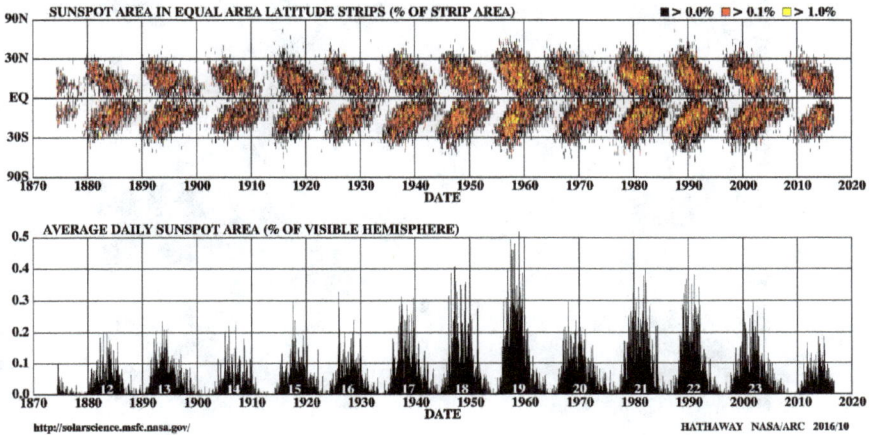

Figure 4.3. *Lower panel*: The average fractional area of the solar disk occupied by sunspots. *Upper panel*: 2D plot of the area occupied by sunspots at different latitudes as a function of time. (Courtesy of NASA/MSFC and Dr. David Hathaway at NASA/Ames Research Center.)

and was surprised to see no cyclonic motion, but rather a radial outward motion from sunspots. This discovery, in 1909, is yet to be satisfactorily explained. Hale's discovery of magnetic fields in sunspots led to a clue about why sunspots are dark, only 50 years later. The science of conducting fluids became fully developed only in the middle of the 20th century, culminating in the discovery of hydromagnetic waves by Hannes Alfven, for which he received the Nobel Prize in 1970. This new branch of physics began to be vigorously applied to the problem of magnetic field production in the Sun. In the regions of strong magnetic fields, the movement of plasma across the fields sets up an inductive electric field, very much like the flow of induced current along a coil of wire that is rapidly moved in a magnetic field. This effect is known as Faraday's law of electromagnetic induction. This induced field prevents the plasma from flowing across the lines of force. We have already seen how energy is transported in the Sun's interior by convection. This convection is prevented in strong magnetic fields, a fact that was discovered by Subrahmanyam Chandrasekhar. When the

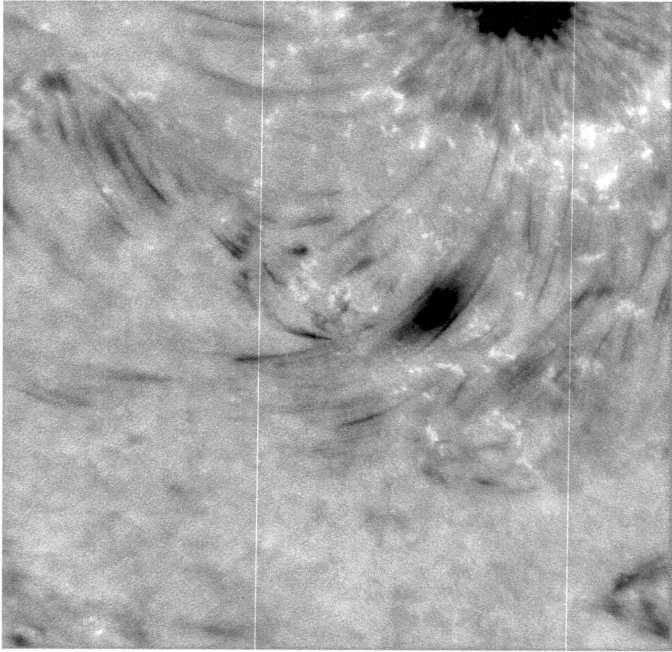

Figure 4.4. Whirlpool-like structure of a sunspot seen in chromosphere light, observed using the Multi Application Solar Telescope of the Udaipur Solar Observatory (USO). (Courtesy of USO/PRL.)

energy transported is blocked, no light will reach the surface of the sunspots, making them appear dark. What about the explosion that can result if energy is blocked at some place? This is still a matter of research right now. One view is that the blocked energy gets around the sunspot zone and appears on the surface at neighboring sites. Another view is that, although convection with overturning motions is prevented, there could be "oscillatory convection," i.e. an up-and-down movement of the plasma, that could store the energy during some phases of the oscillation and dump it at some place far above the photosphere during other phases of the oscillation. The explanation for the 11-year cycle is also a matter that has been pursued very hotly in recent times. New tools for probing deep into the interior of the Sun, and new capabilities of doing simulations of solar convection in a magnetic field using the power

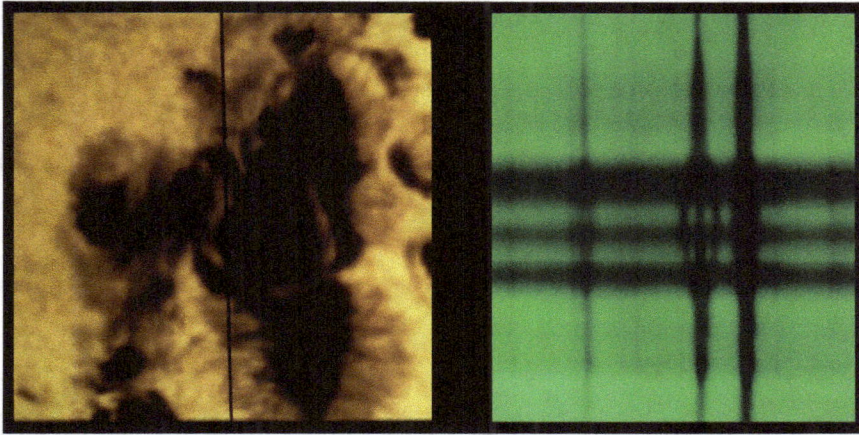

Figure 4.5. The left panel shows the image of a sunspot, with the dark vertical line in the middle showing the position of the entrance slit of the spectrograph. On the right you can see the spectrum of the Zeeman-sensitive Fe lines in the solar spectrum. The dark horizontal bands are at the locations of the sunspot's umbra on the slit, while the dark vertical lines are the spectral lines. Notice the larger splitting in one of the spectral lines. This shows that this particular line is more sensitive to the Zeeman effect compared to the other spectral lines. (Credit: NSO/AURA/NSF.)

of modern behemoths of computing, have started giving a glimpse of what might be. Since these results are very new, it will take a while before this is all confirmed properly. In the meantime, some fundamental properties of the "solar dynamo" that generates the solar magnetic field have been identified, and we will have a look at them.

4.3. The Creation of Solar Magnetic Fields

If we look at a map of the solar magnetic field (Figure 4.6), we will get the impression that the field is distributed in a random and irregular way over the solar surface

However, there are some regular patterns of behavior of the sunspot magnetic fields which were discovered by Hale after several years of patient observations. He noticed that sunspots always appear as pairs with a bipolar magnetic structure. The leading (westward) spot always had a definite magnetic polarity, which

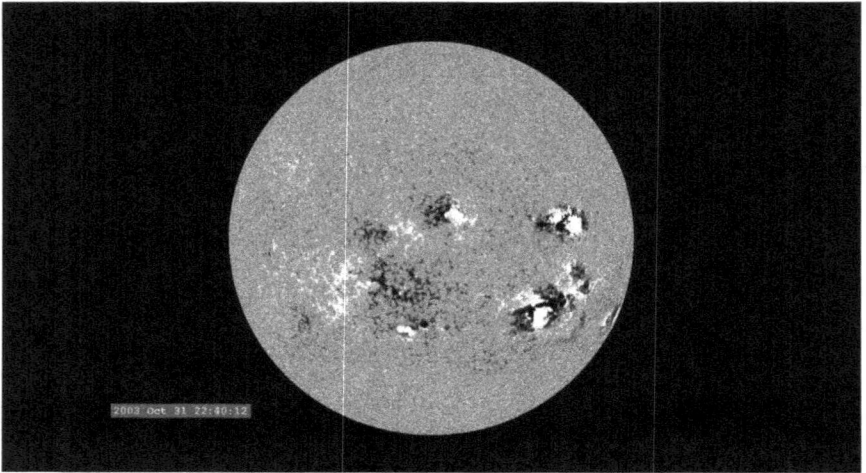

Figure 4.6. The map of the line-of-sight component of the magnetic field is depicted as intensity, the darkest being strong negative (south) magnetic polarity and the brightest being strong positive (north) magnetic polarity. (*Source*: https://svs.gsfc.nasa.gov/3503. Credit: NASA/Goddard Space Flight Center Scientific Visualization Studio.)

was opposite to the following (eastward) spot, during every 11-year sunspot cycle. But the pairs in the northern hemisphere had an opposite pattern of polarity compared to the pairs in the southern hemisphere. The sign of polarity also reversed in every alternate 11-year cycle, which led Hale to the discovery of the 22-year magnetic cycle of sunspots. This type of systematic behavior over a long time clearly indicates a global origin for the magnetic fields of sunspots (see Figure 4.7).

A simple model to explain this pattern of magnetic field changes was suggested by Babcock (Figure 4.8). He first imagined the field to be made up of a poloidal part and a toroidal part. The poloidal part has field lines that remain confined to planes that contain the polar axis on their surfaces. The toroidal part has field lines lying on planes perpendicular to the polar axis. If the Sun rotated about its axis with the same period at all the latitudes, then this poloidal field would merely move round and round in a way without any relative motion between the field lines (remember that the magnetic

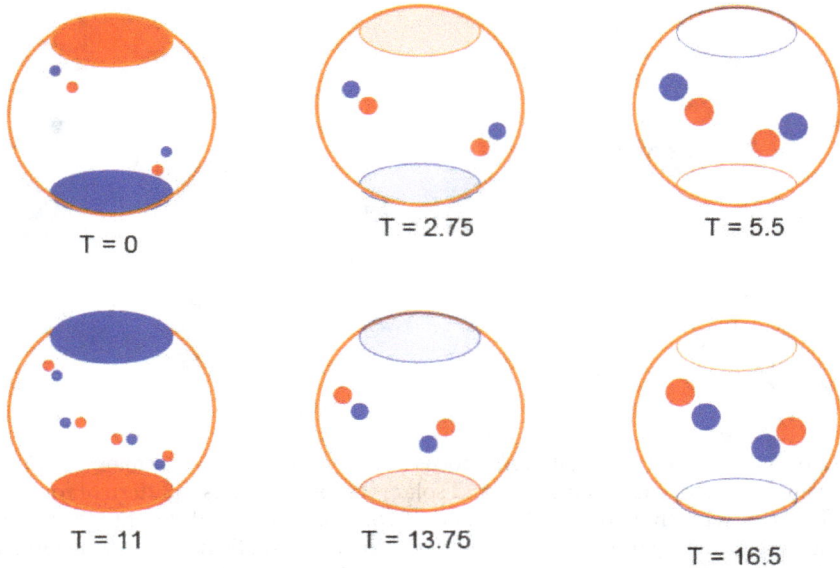

Figure 4.7. Schematic picture of the behavior of the magnetic polarity pattern during a 22-year magnetic cycle. At $T = 0$, it is the beginning of a new cycle with bipolar spots appearing at extreme latitudes. The polar field is strong and is of the same polarity as that of the leading sunspot. As the cycle progresses, the spots become greater in number (depicted by the size of the bipole) and the polar field becomes weaker. At the sunspot maximum ($T = 5.5$ years), the polar field vanishes and the spots are closer to the equator. At the sunspot minimum, the polar field is once again strong, but with reversed polarity. Spots of the new cycle appear at extreme latitudes, but with reversed polarity, while the spots of the old cycle are at the equator. This continues till the next minimum and repeats every 22 years. The polarity patterns are opposite in both the hemispheres.

field is "frozen" into the fluid because of the very high electrical conductivity of the solar plasma). However, the Sun is known to rotate faster at the equator than at the poles. The uneven rotation of the Sun will introduce a relative motion of the field lines and will distort the field. This distortion will twist the poloidal magnetic field of the Sun into a toroidal field, which appears in the shape of a torus similar to that of an inflated tyre tube around the middle latitudes of the Sun. Hence the term "toroidal field." Such a distortion will continue until all of the poloidal field has been converted into the toroidal magnetic field. The toroidal magnetic field has a

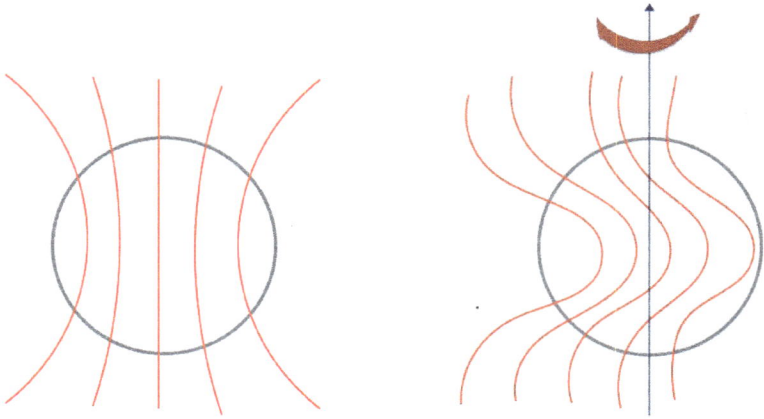

Figure 4.8.   Sketch of the different processes that produce the solar magnetic field. A combination of solar convection and solar rotation produces a pattern of rotation that varies with the latitude. This kind of rotation is called differential rotation. This differential rotation stretches out the poloidal magnetic field (*left panel*) into a belt of magnetic field along the direction of longitude variation (*right panel*).

tendency to become buoyant, very much like a lifebelt underwater. A portion of this toroidal field will float up to the solar surface and pop out through the surface in the form of a bipolar sunspot pair (Figure 4.9). As more and more magnetic field is created in the form of the toroidal field, more and more sunspots appear on the surface. When the strength of the toroidal field decreases, the number of spots also decreases. When all of the poloidal field gets converted into the toroidal field, no more spots will be produced. This is what produces a sunspot minimum.

The sunspot fields are known to fragment into smaller pieces of magnetic flux and decay. There is another behavior of the sunspot bipoles where we always see the line joining the bipoles to be tilted toward the equator. This behavior is known as Joy's law. Thus, when the sunspot field gets fragmented and diffuses out, there is a predominant sign of the field that goes toward the poles and the opposite field diffuses toward the equator. The poleward-propagating field goes to the poles and cancels the existing polar magnetic field. This happens during the sunspot maximum. Thereafter, the polar field attains the opposite magnetic polarity. The tilt produced in

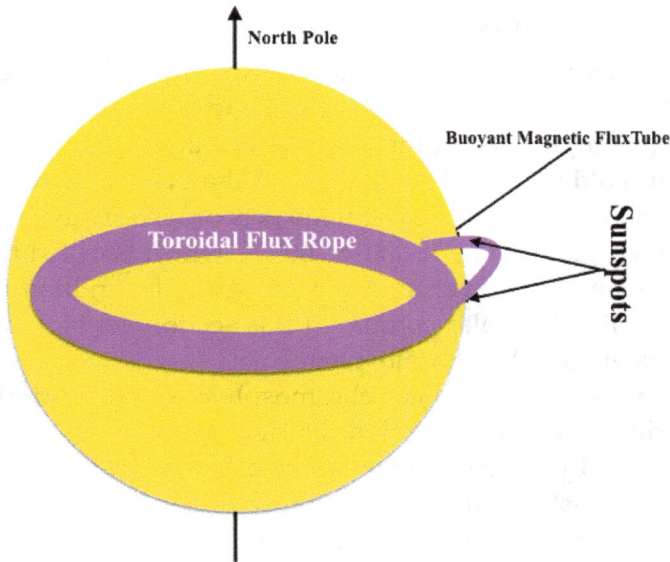

Figure 4.9. The belt of magnetic field produced by the action of differential rotation on a simple poloidal magnetic field can become buoyant and will rise to the surface of the Sun. The cross-section of this popped-out field at the surface will appear as a bipolar magnetic field that harbors the bipolar sunspot group. This figure shows the emerging loop rising through the convection zone through the photosphere.

Joy's law actually converts part of the toroidal field of the sunspots into a poloidal field. It is this poloidal field that ultimately cancels the polar field of the existing sunspot cycle and reverses the sign of the polar field. When the polar field is carried into the interior of the Sun by a meridional flow of the plasma, it becomes the seed poloidal field for the next sunspot cycle. This gets converted once again into the toroidal field and produces the sunspots of the next sunspot cycle. This explanation by Babcock is only a description of the observed changes in the sunspot field using the movements of the solar plasma as a basis. Some simplified dynamo models based on this scenario have been developed and are known as flux transport dynamo models. For a great description of such models, read the book *Nature's Third Cycle,* by Arnab Rai Choudhuri.

4.4. The Solar Chromosphere

In a normal situation, as we go farther away from a hot material, we will feel less and less warm. In the same way, we should normally expect the temperature of the solar plasma above the photosphere to become colder and colder. This is not the case in the solar atmosphere, and the solar plasma actually gets progressively heated up, first to a temperature of 25,000–50,000 K, which is called the chromosphere, and later even to a temperature of a million degrees, which is called the solar corona. In this section, let us look at the properties of the solar chromosphere.

The existence of the solar chromosphere was historically realized during eclipses of the Sun. During a total solar eclipse, the Moon gradually covers more and more of the Sun's disk. A pink flash is seen just before the appearance of the corona (which signals the complete coverage of the disk). This pink flash is the light that comes from a thin layer, about 2000 kilometers across, that is sandwiched between the photosphere and the corona. Actually, it is made up of two parts. The bulk of the layer is at a temperature of a few tens of thousands of degrees, which is hot enough for the major constituents of the atmosphere, such as hydrogen, to start losing the electron from the atom. The recombination of electrons with protons produces, among other emissions, a strong emission of the primary spectral line of the Balmer series of hydrogen. This emission is in the red region of the spectrum and thereby lends its color to the chromosphere. This name originates from chromos ("pink" in Greek). A full disk picture of the chromosphere in the light of the H-alpha spectral line is shown in Figure 4.10. This line can be produced only by hydrogen atoms in an excited state. Excitation of the atom by collision can be possible only if the plasma temperature exceeds 25,000 degrees. A thin layer on top of the chromosphere, called the transition region, has a temperature of 100,000 degrees and mainly emits ultraviolet light.

We can see cloudlike structures suspended above the photosphere. Off the limb, they are bright in H-alpha and are called prominences (see inset of Figures 4.10 and 4.12). On the disk of the Sun, they appear like dark long threads (Figures 4.10 and 4.13)

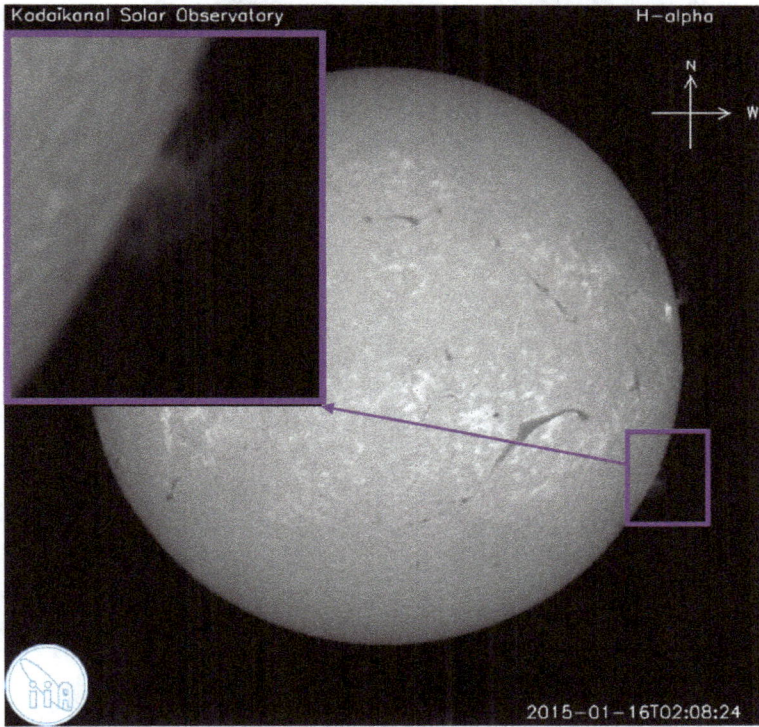

Figure 4.10. The solar chromosphere seen in the light emitted by excited hydrogen atoms. The inset shows a prominence seen off the solar limb. Dark, threadlike features are called filaments and are the projection of prominences on the solar disk. (Courtesy of Indian Institute of Astrophysics.)

and are called filaments. We also see many other fine features in the chromosphere, like spicules, which are tiny jets of plasma moving up with an average speed of about 20 km/s (Figure 4.11). Spicules are better seen off the outer edge of the Sun. No one has clearly identified what structures represent the spicules on the face of the Sun, but a good guess might be the so-called H-alpha fibrils or calcium mottles.

The heating of the chromosphere is a lot easier to understand compared to the heating of the corona. Deep in the convection zone of the Sun, the bubbling convection produces sound waves, much like the splashing sound produced by boiling water. The frequency

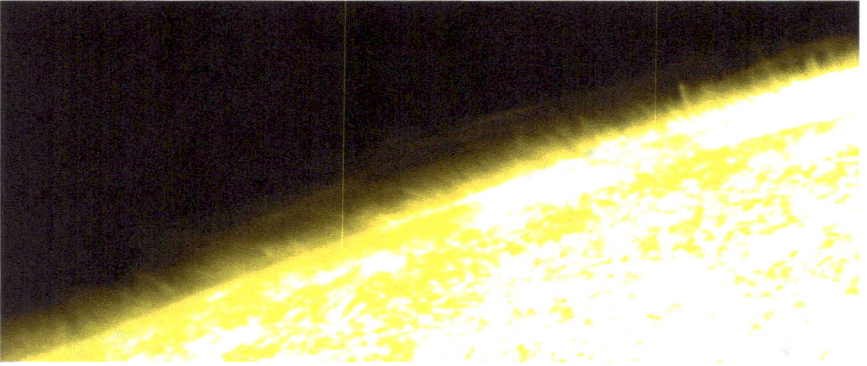

Figure 4.11.   Solar spicules seen as grasslike structures at the edge of the Sun. (Courtesy of NASA/JAXA/Hinode.)

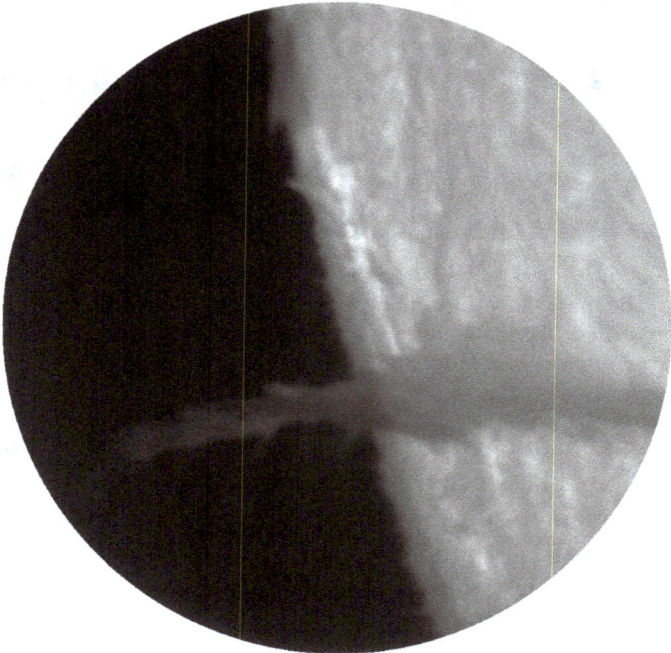

Figure 4.12.   Image of a solar prominence obtained by the Multi Application Solar Telescope (MAST) of the Udaipur Solar Observatory and seen beyond the solar limb. (Courtesy of Udaipur Solar Observatory.)

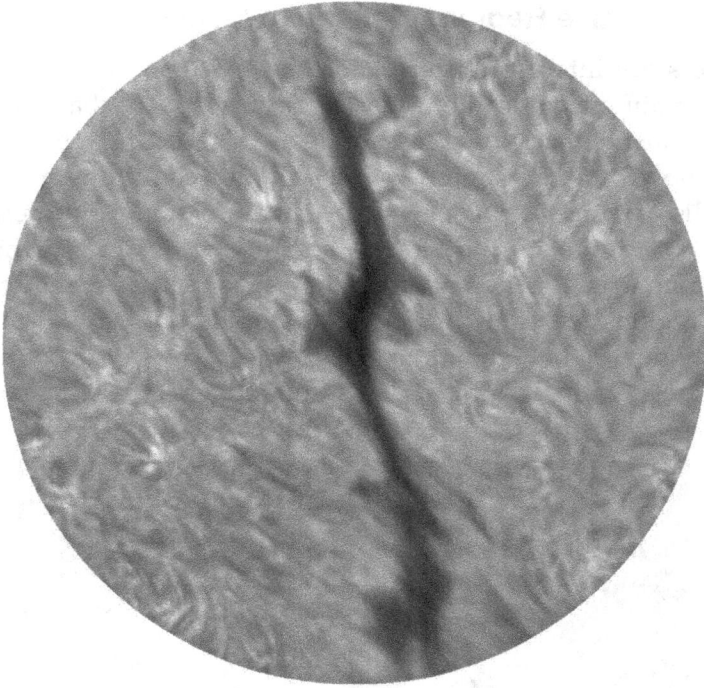

Figure 4.13.   Image of a solar filament obtained by MAST and seen on the solar disk. (Courtesy of Udaipur Solar Observatory.)

of these solar sound waves is in the range of 1/10 to 1/1000 of a cycle per second. In comparison, the human voice has a range of 100 to 1000 cycles per second. When the solar sound waves travel up through the photosphere, they encounter regions of smaller and smaller density. As a result, the waves grow in strength, similar to the increase in the heights of ocean waves as they approach the shallower depth of water near a beach. The ocean waves become very steep before they crash on to the land in the form of foam and surf. In the same way, the solar sound waves steepen and form shock waves. These shock waves heat up the plasma and produce the chromosphere. The magnetic fields seem to be able to increase the amount of chromospheric heating. How they are able to do this is still an open question.

4.5. The Active Regions Related to Sunspots

Sunspots are interesting things in their own right, but they also have certain effects in the upper solar atmosphere that affect our own atmosphere, sometimes seriously. These phenomena have been known to cause interruption in radio communications and even disruptions in power supply in the extreme northern and southern latitudes of the Earth. The Sun's atmosphere above the photosphere in the neighborhood of sunspots is extremely hot. These regions are called active regions. The hot plasma in these active regions is hot enough to emit x-rays. X-rays of the Sun show looplike structures stretching across the active regions (Figure 4.14).

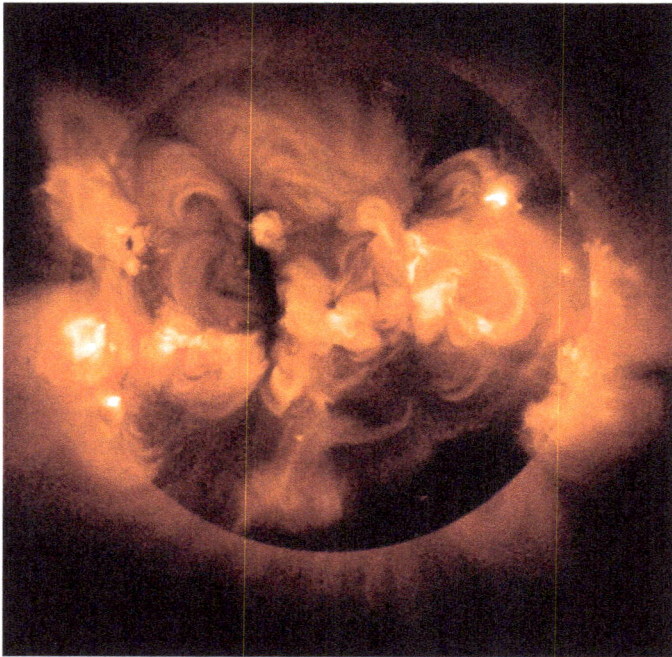

Figure 4.14.   Picture of the x-ray emission of the Sun taken by the Japanese satellite *Yohkoh*. Note the bright, looplike structures at the active regions. There are also dark patches of depleted coronal material which are called coronal holes. These coronal holes are the sources of high-speed streams of plasma coming from the Sun. (*Source*: https://solarscience.msfc.nasa.gov/corona.shtml. Courtesy of NASA/MSFC.)

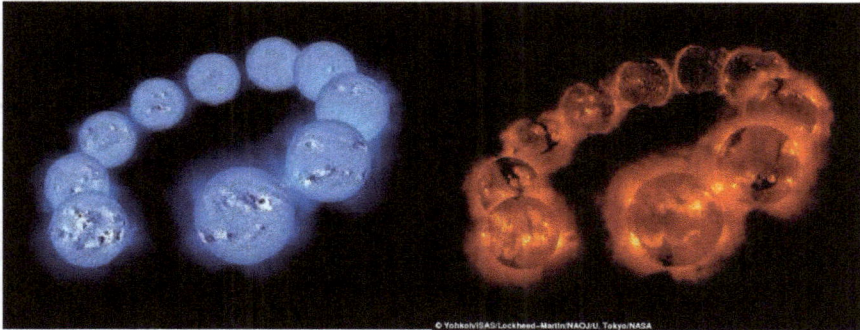

Figure 4.15. *Left*: A sequence of magnetograms at different phases of the solar cycle (https://courses.lumenlearning.com/astronomy/chapter/the-solar-cycle). *Right*: X-ray images for an 11-year solar cycle (https://en.wikipedia.org/wiki/Solar_cycle).

The temperature of the plasma that can emit x-rays is a few million degrees. In comparison, the temperature of the photosphere is only 6000 degrees. Heat always flows from a hotter region to a cooler region. Therefore, the heat must flow from the million-degree plasma down to the photosphere. The time it takes to cool the active regions by this flow of heat is only a fraction of a day. However, the active regions and their loops are known to last for up to a month, and sometimes longer. Clearly, some process is continuously supplying the heat to the million-degree plasma to replace the heat being conducted down to the photosphere. However, what this process is, is anybody's guess. What we do know is that the magnetic field of the sunspots somehow manages to heat the plasma.

The plasma outside of the x-ray loops also has a temperature of a million degrees, though not as hot as the plasma within the loops. Curiously enough, the underlying photosphere, outside of sunspots, is also threaded by tiny magnetic structures, which are hardly 100 kilometers across in size. Therefore, the magnetic cause for the heating of the upper solar atmosphere is now beyond question (Figure 4.15). This hot plasma, above active regions, and elsewhere is called the solar corona.

4.6. The Solar Corona

The existence of the solar corona was known much before x-ray pictures of the Sun were available. During a total solar eclipse, the Moon completely covers the face of the Sun. Those who have seen a total solar eclipse can never forget the awesome scene of the Moon's shadow seen against the brilliant, steel-gray halo surrounding it with magnificent streamers reaching out from the Sun radially in all directions (Figure 4.16).

This halo or crown gave the name "corona" to this gas. Why do we not see the corona every day? The culprit is the molecules of our own atmosphere, which gives the color blue to our skies. These sky molecules scatter at least one thousandth of sunlight toward us. The coronal material is so tenuous that it scatters only a millionth of the sunlight in the same direction. So, this feeble light is more than swamped by the skylight, preventing us from getting a daily view of the corona. It is only when the Moon shuts off the

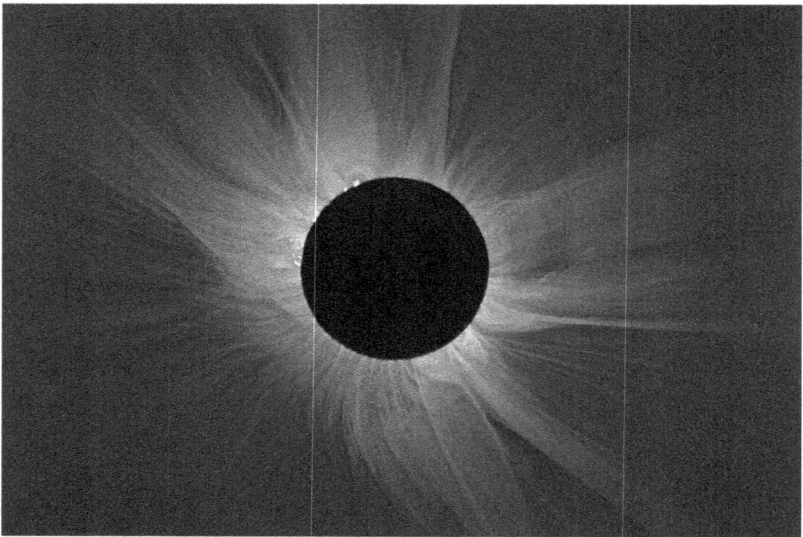

Figure 4.16. Picture of a total solar eclipse showing the solar corona (https://solarscience.msfc.nasa.gov/corona.shtml). (Credit: NASA/MSFC and HAO/NCAR.)

bulk of the sunlight that the corona is able to shine through, in the near-nightlike conditions. The x-rays emitted by the corona are stronger than the x-rays present in normal sunlight. Fortunately, for all living creatures on the Earth, these photons are absorbed by our atmosphere. Thus, we need to go to outer space to watch these x-rays. Indeed, the astronauts manning the satellite *Skylab* did just that and beamed back excellent pictures of the x-ray corona. More recently, the Japanese satellite *Yohkoh* (meaning "sunbeam" in Japanese) has collected an unprecedented number of x-ray pictures of the corona (Figure 4.14). Even more recently, the satellite *SOHO* has started collecting data about the corona, which can perhaps solve the vexing problem of what keeps the corona so hot. The most recent coronal images have been obtained by the satellite *SDO* (*Solar Dynamics Observatory*). The reader can see these beautiful pictures on the website https://sdo.gsfc.nasa.gov.

Historically, even before x-ray pictures of the corona were available, astronomers analyzed the visible light during solar eclipses and found spectral lines that did not seem to fit into the pattern of spectra emitted by all the chemical elements. For a while, there was talk about new elements. Later on, some excellent bit of scientific reasoning using the spectra from iron arcs led the spectroscopists, such as Edlen, to identify the coronal emission as coming from an iron atom that has been stripped of 13 electrons. Now, it is not easy to knock away so many electrons from the iron atom unless it is immersed in a very hot gas. Thus, astronomers had to conclude that the corona was very hot — a few million degrees hot, to be precise. Eclipses also provided a clue as to what could be the agent that heats the corona. Eclipses that occurred when the Sun had a large number of sunspots showed a very bright corona. Those that occurred during a lean phase of the sunspot cycle, when very few sunspots are present, showed a weaker coronal brightness. Since more sunspots means stronger magnetic fields, it is not difficult to conclude that the magnetic field of the Sun plays an important role in the heating of the solar corona. Alas, in spite of a large number of attempts, there is no understanding yet of how exactly the magnetic field heats the corona. When the understanding dawns, it will not only solve a vexing solar puzzle, but might also give us a clue as to how to maintain million-degree plasma in devices called tokomaks. This last is an important thing, because tokomaks can be tomorrow's solution to the growing energy crisis on the Earth.

4.7. The Solar Wind

I have mentioned earlier that the Moon lost its atmosphere because the Moon's gravity is not strong enough to pull down the energetic molecules of the atmosphere, which gets heated by sunshine. In comparison, the particles of the coronal plasma will be extremely energetic at the temperature of a million degrees. The average speed of their to-and-fro motions is about a hundred kilometers per second. Close to the solar surface, the Sun's gravity is strong enough to retain all these energetic particles. However, the force of gravity decreases with distance from the solar surface. Beyond a particular distance, the gravity will no longer be strong enough to contain the hot plasma. The plasma will start escaping from the Sun at this point. This steady leak of the corona continues at greater and greater speeds, until the stream of particles develops into a wind, which blows at supersonic speeds. The German astronomer Bierman noticed that a comet's ionic tail always pointed in a direction along the line connecting the Sun to the comet (Figure 4.17). This made him think that there is steady wind from the Sun that creates a deflection of the cometary tail.

The solar physicist Eugene Parker predicted the existence of the wind and estimated that it would reach the Earth with a speed of a few hundred kilometers per second. At that time, it was a rather bold prediction and his scientific paper on this prediction was rejected by referees. The then editor of *The Astrophysical Journal*, Subrahmanyan Chandrasekhar, overruled the referees and published the paper. Later on, with the advent of the Space Age, instruments aboard satellites indeed detected this wind. This was a real triumph for theoretical solar physics. Years later, it was noticed that the solar wind was very "gusty," containing very-high-speed streams mixed with slower winds. These high-speed streams rotate with the Sun like a searchlight beam, and caress the Earth every 27

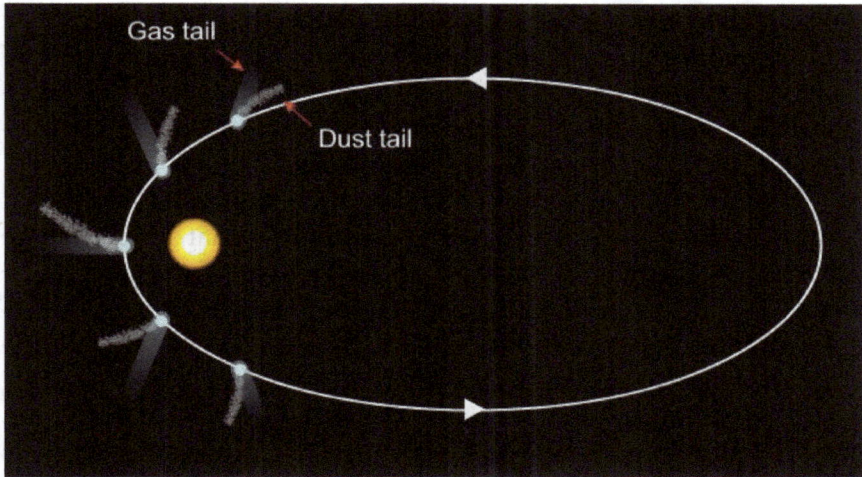

Figure 4.17. Deflection of the ionic tail of a comet from its dusty tail, showing evidence of the solar wind (https://upload.wikimedia.org/wikipedia/commons/8/81/Cometorbit01.svg).

days. They seem to originate from large regions of single magnetic polarity on the Sun. These regions are different from sunspot fields, which are bipolar and compact. The unipolar magnetic regions also appeared as very dark regions in the x-ray pictures, and gained the name "coronal holes." But the physical explanation for the acceleration of the solar wind to speeds much beyond the value predicted by Parker still eludes the scientists.

4.8. The Solar Oscillations

You might have perhaps got the impression by now that the Sun is interesting only because of the activity going on at various heights of the solar atmosphere. However, even outside of the active regions, the Sun is not completely still. Scientists discovered that these so-called "quiet" regions of the Sun are actually moving up and down in regular patterns. The up-and-down movement also has a rhythm in time. The basic period of this rhythmic motion is about five minutes. Initially, when these oscillations were discovered, they were treated as a mere curiosity. What intrigued the scientists was the regular nature of these oscillations, which were uniformly present all over the Sun, with a definite range of sizes. As usually happens in science, a few people started wondering about these oscillations. The answer had far-reaching implications for not only astronomy but also physics!

To understand this, we will have to learn something about sounds emitted by various objects. You know that long or large musical instruments produce a deep bass sound, and that short or small instruments produce a shrill sound. Thus, the prominent frequency emitted by these instruments depends on the size of the emitter. When you pluck a guitar string, you are actually exciting a very wide range of frequencies in the string. However, the string, in turn, will respond only to a few definite frequencies. The periods corresponding to these frequencies are usually some multiples of the time interval taken for a sound wave to travel from one end of the string to the other. The longer the string is, the longer it will take for sound to travel, and the larger will be the dominant period of the emitted sound. You therefore end up producing a bass sound. The dominant frequency is also the natural frequency of the string. Every mechanical structure has its own set of natural frequencies. When an earthquake strikes a building, the building will shake only at its natural frequencies. If the earthquake waves also have a large power in some of these frequencies, then the response will be all the more violent and the building is very likely to break up.

The Sun is also a mechanical structure, held together as it is by a balance between pressure forces and gravity. It will also have its natural frequencies (Figure 4.18). What is more, each of these natural frequencies has an associated pattern of oscillations that has a maximum at a particular depth of the Sun. For example, the lower frequencies (the bass sounds) will have a maximum response at deeper layers of the Sun, while the higher frequencies will respond better at the outer layers of the Sun. Now, we already know that the convection zone produces a lot of sound waves, but in a haphazard way, as is common for convecting or boiling fluids. The period of five minutes happens to have the greatest response in

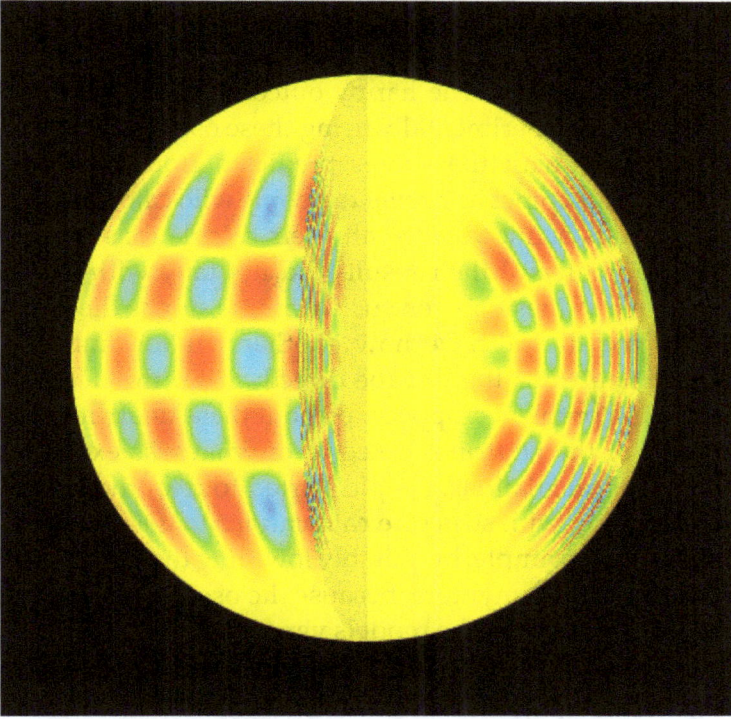

Figure 4.18. Some possible modes of oscillations of a spherical mechanical object like the Sun. Red depicts the red-shifted region with motion away from observer. The real sun consists of millions of such modes. (https://en.wikipedia.org/wiki/Helioseismology)

the convection zone. It is thus that the dominant period of solar oscillations happens to be about five minutes. The Sun responds maximally with a natural oscillation period of five minutes, which it picks out from among the jumble of frequencies generated by the convective motions.

The story of the natural or global oscillations of the Sun does not end here. As soon as the scientists realized that the five-minute oscillation is the natural oscillation of the Sun, they started devising clever ways of using these oscillations to probe the properties of solar material in the very interior of the Sun. By analyzing the arrangement of the frequencies and the size of the oscillating patches associated with each of these frequencies, the scientists are now able to tell us the variation of the sound speed at different layers of the solar interior. The happy outcome of all this is that the values measured experimentally using these oscillation frequencies tally very nicely with the values predicted from the theoretical calculations. This comes as a relief to people who calculate the structures of various stars, because all their calculations would have become meaningless if their results did not tally with experimentally measured values for even such a simple star as the Sun.

This is not all. We have now another window into the Sun, namely the neutrinos that escape directly from the deep interior. A most vexing problem cropped up about these neutrinos. Initially, only a third of the neutrinos detected of what was expected from the theoretical calculations. People started to treat the lack of neutrinos by fiddling with structure calculations to get a lower output of neutrinos (for example, by suitably lowering the central temperature of the Sun). Not anymore, because the oscillations tell you that you cannot tamper with the models very much. Thus, the neutrino puzzle became a serious puzzle and particle physicists came up with ingenious explanations for the lack of neutrinos. These explanations do not meddle with the Sun; rather, they meddle with the neutrinos themselves. The mystery of the missing neutrinos was eventually solved when all flavors of the neutrinos were measured in contrast to the early experiments, which measured only one flavor. The expected neutrino count was finally seen when all flavors were accounted for. The neutrinos had been naughtily switching between flavors when they traveled from the Sun to the Earth and thereby created a scare. More details will be given in the last chapter about neutrino measurement.

5 THE SPHERE OF INFLUENCE

5.1. Introduction

We all know that the Sun exerts a strong influence on terrestrial weather and climate. The changing seasons, the prevailing winds, the ocean currents — all these important ingredients that make up our weather and climate depend very strongly on the solar energy output and its variations with time. What we do not generally know is the fact that solar activity, namely flares, solar mass ejections, and variations in solar ultraviolet emissions also causes significant changes on the Earth's atmosphere (see Figure 5.1). However, the changes wrought on the upper atmosphere have a profound influence on high-technology systems that form a part of modern life. They also affect strategic maneuvers of modern defence systems. Thus, the subject of space weather (as this branch of solar — terrestrial physics is now called) has gained both commercial and military importance in recent times. In what follows we will see how each manifestation of solar activity has a particular effect on space weather

5.2. Solar Flares

The hot coronal loops might appear somewhat steady in x-ray "still" pictures. If we were to keep on watching these loops, though, we would soon get a different impression altogether. Sometimes, we would be able to catch a loop that was preparing to produce an explosive phenomenon called a solar flare. The loop would get heated up, thus appearing to brighten up in x-rays. Then, after

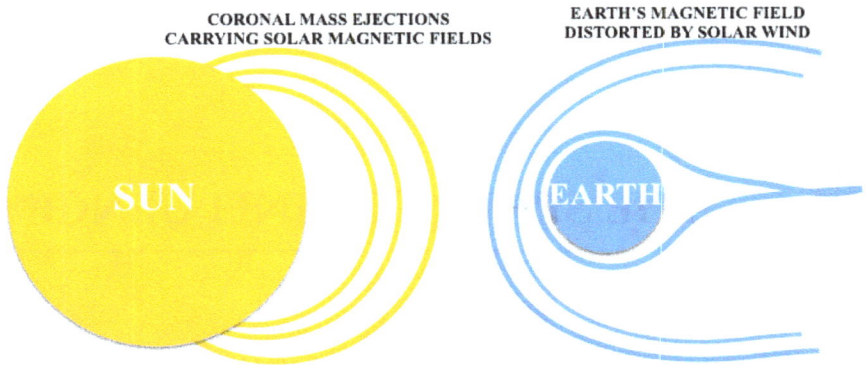

Figure 5.1.   Schematic of the interaction of Sun and Earth.

about 15 minutes or so, several things would start happening all at once. The loop would suddenly break open spewing out a clump of hot plasma (plasma made up of energetic particles like electrons, protons, and alpha particles). Some of the electrons would travel down along the legs of the loop, heating the denser gas at the feet of the loops. When striking the denser matter, the electrons, would emit x-rays, radio waves, and sometimes gamma rays (Figure 5.2). The heated matter would also expand upward, filling the entire loop with the hot plasma. This would be seen as an enormous brightening of the loop in x-rays. The cloud of gas that was ejected initially would push out the matter ahead of the cloud, forming a shock wave. This shock wave would further energize the particles of the cloud, producing particles of incredibly high energies. These energetic particles would then shoot away from the Sun. Some of them would hit the Earth's protective magnetic field, producing magnetic storms that could even disrupt the communication systems on the Earth. The x-rays and gamma rays would arrive much earlier (because they travel at the speed of light), and ionize the neutral atoms in the Earth's upper atmosphere. The electron content of the ionosphere would increase, causing shortwave fadeouts in radio broadcasting.

For those who do not perch on satellites, a solar telescope on the ground will show a sudden brightening in the light of hydrogen

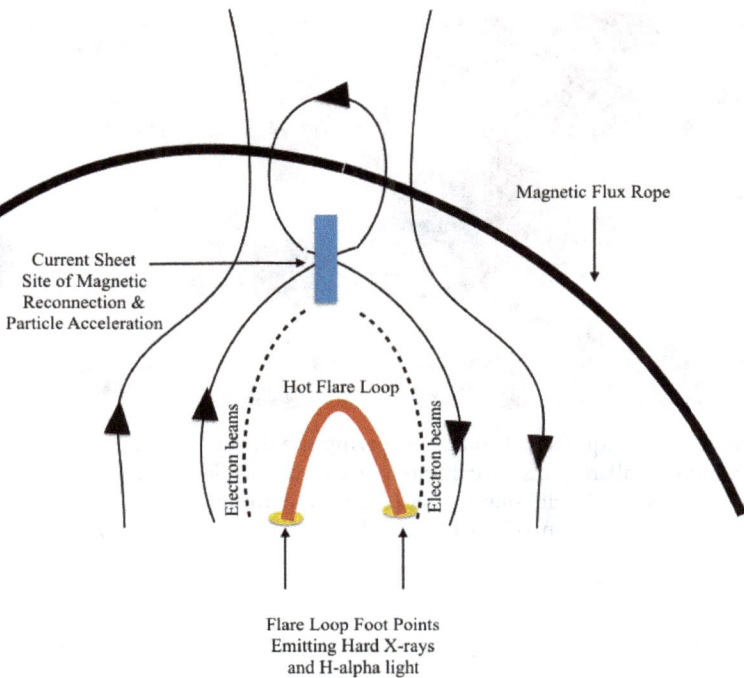

Figure 5.2. Schematic of the solar flare process. A current sheet forms when two oppositely directed field lines are pushed together. Reconnection at this site results in particle acceleration. The accelerated particles travel down to the chromosphere, which acts as a thick target, producing hard x-rays and H-alpha emission. The target plasma heats up and evaporates, filling the magnetic loops with hot plasma, which emits soft x-rays.

alpha. Two ribbons of intense light will appear at about the time that the electrons coming down the loop hit the denser material at the feet of the loop. These ribbons will then move apart for a few minutes, reach a maximum separation, and then fade away gradually (Figure 5.3). The sudden brightening at the onset of the flare will be accompanied by a pulse of microwave radiation that can be detected with radio telescopes.

What exactly causes a flare is not known yet. The magnetic fields at the base of the loops that produce the flares are seen to be in a very twisted and stressed condition before the flare. It is now believed that the twisted and stressed fields will violently relax to

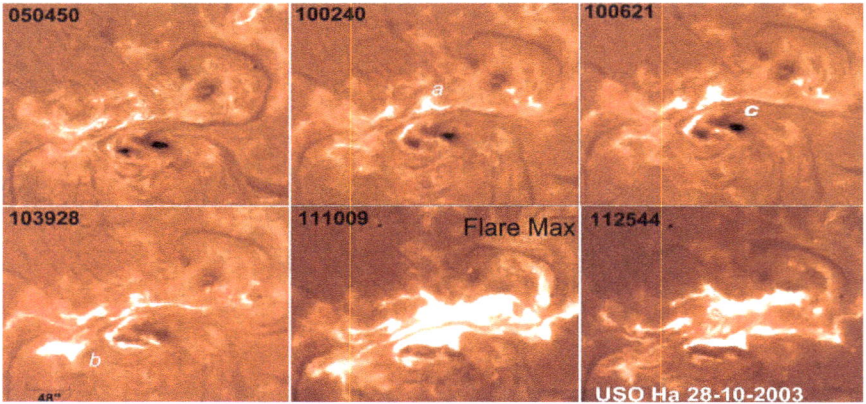

Figure 5.3. Sequence of images showing the development of a solar flare. Starting with small regions of brightening called flare kernels, the flare proceeds in the form of bright ribbons which reach a maximum and then decay back to the initial quiet condition. (Courtesy of Udaipur Solar Observatory/Physical Research Laboratory.)

a stress-free state whenever they are twisted beyond a certain limit. This relaxation possibly provides the energy that is released explosively in a flare. However, there has been no clear-cut evidence for this field relaxation. Flares have not yet been caught in the act. If we were to fully understand the mechanism that produces a flare, then we might even be able to predict when exactly a flare is going to erupt. This capability of flare prediction will be a great boon to humankind.

Historically, Carrington first proposed a terrestrial effect of a solar flare. He noticed that intense geomagnetic storms occurred soon after a large white light flare erupted on the solar disk. However, he was cautious about relating the two phenomena and remarked, "One swallow does not a summer make." Lord Kelvin, on the other hand, discounted any possibility of a physical connection between the two phenomena and went on to prove his point by showing that the induced magnetic field produced by the flare at a distance of 1 AU, was insignificantly small. His calculations were impeccable, but he had assumed a vacuum between the Sun and the Earth. What we do know now, of course, is that solar energetic particles and shocks propagate in the interplanetary medium and produce geomagnetic storms. Solar flares, in fact, emit a wide variety of radiations, and an equally wide range of effects are produced in our atmosphere.

5.2.1. The effect of solar x-ray and extreme ultraviolet flux

Under normal conditions, solar x-ray and extreme ultraviolet radiations are absorbed in the Earth's thermosphere (100 kilometers above ground) and ionize (remove electrons from) the atoms of the thermosphere. This process leaves a layer of free electrons and ions, called the ionosphere. Shortwave radio communications make use of this ionosphere by bouncing the radio waves back to the ground, so one can connect large distances by radio.

The number density of electrons decides the frequency of the selectively reflected radio wave. The larger the density, the shorter the wavelength of the reflected wave. Now, solar x-ray and extreme ultraviolet radiations increase by large amounts during a flare. This causes a temporary increase in the number density of electrons at different heights. Two stations, under perfect communication during normal conditions using a particular band of radio waves, will suddenly get cut off because the height from which the radio waves get reflected will suddenly change. This phenomenon is called a shortwave fadeout.

Modern communication systems depend on satellites for bouncing the signals back to the ground. These waves will have a definite polarization, which means that the electric fields of the radio waves always remain in a particular plane. The increased radiation during a solar flare can cause a change in the plane of polarization of the radio waves used in communication. This will produce a distortion in the received signal and thus affect the communications between a satellite and a ground receiving station. Sudden changes in the ionosphere induced by solar flares can also affect the accuracy of the positions determined by the GPS (Global Positioning System). The GPS uses a combination of satellites and provides the latitude and longitude of any desired location on the

Earth with a good accuracy. It facilitates navigation at sea, in the air, and even in modern automobiles by locating the car on a map of a city and providing information to the driver to reach a desired destination. Therefore, you can now see that providing accurate GPS positions has commercial importance. The GPS is now trying to be less vulnerable to flares by using fixed ground-based reference locations and comparing the measurements at more than one frequency. This increases the cost, and single frequency GPS users will find it more economical to know whether any flare is going on so that they can rely on the positions only during flare-free intervals. On a longer timescale, the ultraviolet emissions make the upper atmosphere expand and increase the drag on low-Earth-orbiting satellites. This causes errors in predicting satellite orbits. There is an interesting story regarding this. NASA always relies on long-term forecasts of sunspot number to predict the drag on satellites. The solar cycle 21 (1975–1986; maximum at 1980) had an unprecedented rapid rise to the maximum which resulted in an unexpected large drag on the satellite *Skylab* (1973–1979), which ironically was used for observing the Sun. Plans were made to refurbish and reuse *Skylab* by using the Space Shuttle to boost its orbit and repair it. However, due to delays in the development of the Space Shuttle, *Skylab*'s decaying orbit could not be stopped. *Skylab* eventually fell on an uninhabited stretch of land in Perth, Australia.

5.2.2. Solar radio flux

During a solar flare, the radio emissions from the Sun are enhanced in a very broad band of the radio wave spectrum. This will create interference in a wide variety of terrestrial communications, such as spacecraft-to-ground signals, search-and-rescue signals, and communication with civil and military aircraft.

5.2.3. Solar energetic particle events

A large amount of particles like electrons, protons, and alpha particles are accelerated during a solar flare. They achieve high energies and shoot out from the Sun toward the Earth. Very energetic particles travel at high speeds and reach the Earth within 10 hours of the occurrence of a solar flare. The Earth's magnetic field acts as a protective shield against charged particles during normal times. However, during flares, the flux level and energies are so high that the shielding does not work very effectively, especially at the extreme north and south latitudes. The effects are more severe in outer space. The collision of an energetic particle with high-density electronic circuitry can cause unwanted changes in the operation of spacecraft. People on space missions are exposed to a higher level of danger from radiation. Radiation can also cause permanent damage to spacecraft components and to solar panels used in generating power for the spacecraft.

5.3. Coronal Mass Ejections

Solar flares are spectacular events and have always stolen the show in the context of solar–terrestrial effects until recently. However, solar physicists have come to recognize another phenomenon to be of equal, if not greater, importance: coronal mass ejections (CMEs), which were first discovered by the *Solar Maximum Mission* satellite. CMEs are the eruptions of large coronal helmet streamers. The explosive energy release in solar flares was initially considered to drive the eruptions. Now, we recognize that CMEs are eruptive events in their own right (Figure 5.4). They are often associated with flares, but several events have occurred without flares. CMEs propagate as interplanetary disturbances and have a 50% chance of producing a geomagnetic storm when they reach the Earth. In other words, one out of every two ejections that reach the Earth produces a geomagnetic storm.

Solar flares and CMEs both involve violent release of energy, but are otherwise quite different. The strongest flares are quite often associated with CMEs — but they spew out different things, and they have different effects near planets.

Both eruptions are powered by the sudden release of magnetic stress that builds up by the process of the twisting of magnetic fields. Both eruptions are triggered by the process called magnetic reconnection, which involves the cutting of magnetic field

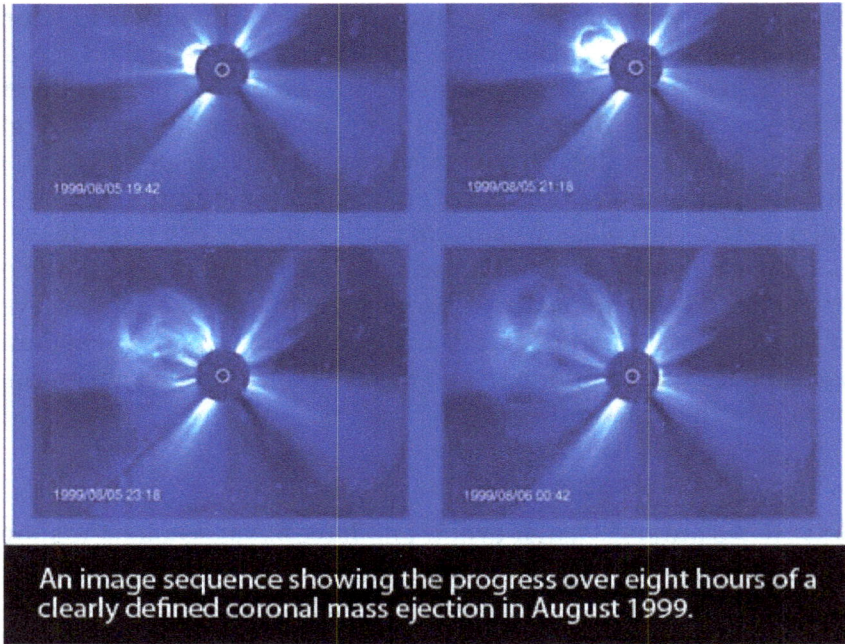

An image sequence showing the progress over eight hours of a clearly defined coronal mass ejection in August 1999.

Figure 5.4. Sequence of images of the solar corona obtained from a corona-graph called LASCO-C3 on board the *Solar Heliospheric Observatory (SOHO)*, a joint project of the European Space Agency (ESA) and the National Aeronautical Space Agency (NASA). We can see a large coronal mass ejection (CME) emerging from the southwest limb of the Sun and spreading out into a tenuous plasma as time progresses. (Image credit: NASA SOHO.)

lines. As mentioned earlier, flares release accelerated energetic particles and high-energy photons. But these travel extremely fast and reach the Earth in several tens of minutes (photons reach it in about eight minutes).

The magnetic stress can also create a CME, most probably under the influence of a large untwisting motion of a previously twisted magnetic field that throws out bulk plasma into space. This plasma carries the magnetic field which is "frozen" with the solar coronal material on account of its high electrical conductivity. Recently, scientists have been able to track these ejections out to large distances from the Sun. They found that, rather than being passively carried along the interplanetary matter riding on the ever-flowing solar wind, these ejections seem to have a motive force of their own and are even accelerated at the further distances from the Sun. Put very simply, the differences between the two types of eruptions can be seen directly through solar telescopes, with flares appearing as an intense emission of radiant energy and CMEs appearing as enormous balloons of gas hurtling into space.

The impact of a CME on the Earth's magnetosphere can also trigger plasma processes that lead to prolonged geomagnetic disturbances. Small changes in the Earth's magnetic field acting over large lengths can induce currents in power distribution systems on the Earth. This introduces a direct current that can sometimes shift the neutral level of the distributed AC power beyond the operating range of the power transformers and cause serious disruption of the power distribution. This problem is most severe for power companies operating at extreme latitudes. An eight-hour collapse occurred in the power distribution system of Quebec Hydroelectric during the magnetic storm of March 1989, and the combined cost of power loss in the United States and Canada was more than US$100 million. The large scale of the CME-induced geomagnetic disturbances creates problems for oil pipelines carrying oil over large distances. Pipeline companies add a bias current to the pipelines to prevent corrosion. This bias current must be set in the absence of large-scale currents induced in the pipelines by geomagnetic disturbances. Thus, the pipeline companies need to know the time, duration, and level of geomagnetic disturbances so that they can avoid setting their bias currents during these disturbed conditions. The charged particles from a geomagnetic storm excite the atoms in the polar regions of the Earth, since they enter the Earth's atmosphere along the polar magnetic field lines. This produces glowing gases which appear as aurorae at the extreme latitudes on the Earth (Figure 5.5).

Geomagnetic storms dissipate the electromagnetic energy by resistive heating of the plasma. This makes the plasma expand and increase the density levels of low Earth orbits. For large events, this can increase the drag as much as to cause errors in orbit prediction

Figure 5.5.   The charged particles from a geomagnetic storm excite the atoms in the polar regions of the Earth, since they enter the Earth's atmosphere along the polar magnetic field lines. This produces glowing gases which appear as aurorae at the extreme latitudes on the Earth. This is a picture taken from the ground. (*Source*: http://wallup.net/nature-landscape-aurorae-2)

and loss of knowledge of satellite locations. In fact, during the same magnetic storm of March 1989 mentioned earlier, more than 2000 objects tracked by the United States Air Force were temporarily lost.

Geomagnetic storms increase the flux of low-energy electrons and protons. This is apart from the direct injection of high-energy particles from the solar flare. This charges the surfaces of space-craft. The sunlit surface loses charge by combining with pho-toelectric charges. The shaded surfaces remain charged. Thus, a voltage can develop on the surface. This sometimes results in an electric discharge that can damage sensitive instruments on the spacecraft.

The Earth's magnetic field provides a good navigational aid. Some geophysical exploration systems also use the Earth's magnetic field as a reference direction. Drilling heads use such a reference. Changes in the Earth's magnetic field will affect all these systems. Naturalists believe that some migratory birds use the Earth's magnetic field as a reference. A geomagnetic disturbance during the bird's flight would make it lose track and land in unfamiliar wintering locations. This would not have been a great calamity for the birds in the past when ecologically favorable locations were aplenty. In recent times, conscious efforts have been made to preserve certain locations to welcome migratory birds in winter. The Bharatpur sanctuary in North India is one such location. However, for dwindling numbers of certain species (like Siberian cranes), losing track during a single flight might prove disastrous to the species, if the birds happen to land in unfriendly locations. The naturalists must keep in mind this possible solar–terrestrial effect on the survival of rare species of migratory birds.

The magnetic field in the CMEs has to face southward to produce a serious effect on the Earth. Since geomagnetic disturbances cause many problems on the Earth, it is now necessary to watch out for the CMEs. Unfortunately, these are not as easy to detect as flares, since either they require special telescopes called coronagraphs kept on high mountains, or they require satellite-borne optical coronagraphs and x-ray telescopes. CMEs can also be seen with low-frequency radio telescopes. Keeping in mind the various kinds of damage caused by geomagnetic storms (mentioned earlier), it would seem necessary to have several such radio telescopes operating at different longitudes so that the Sun can be monitored continuously. The radio telescope at Gauribidanur in South India, operating at 35 MHz, and the one at Nancay, France, operating at 160 MHz, seem now to be the only dedicated solar radio telescopes at low frequencies, the rest being closed down due to radio interference at the frequencies at which mobile phones communicate with each other!

The study of CMEs, including their physical origins and capability for prediction, has resulted in a new branch of solar physics called heliophysics. In recognition of the growing importance of heliophysics for human activities, the year 2007 was declared as the International Heliophysical Year, which coincided with the 50th anniversary of the International Geophysical Year (1957). This marked a remarkable realization of the importance of the effects of solar activity on the Earth (Figure 5.6).

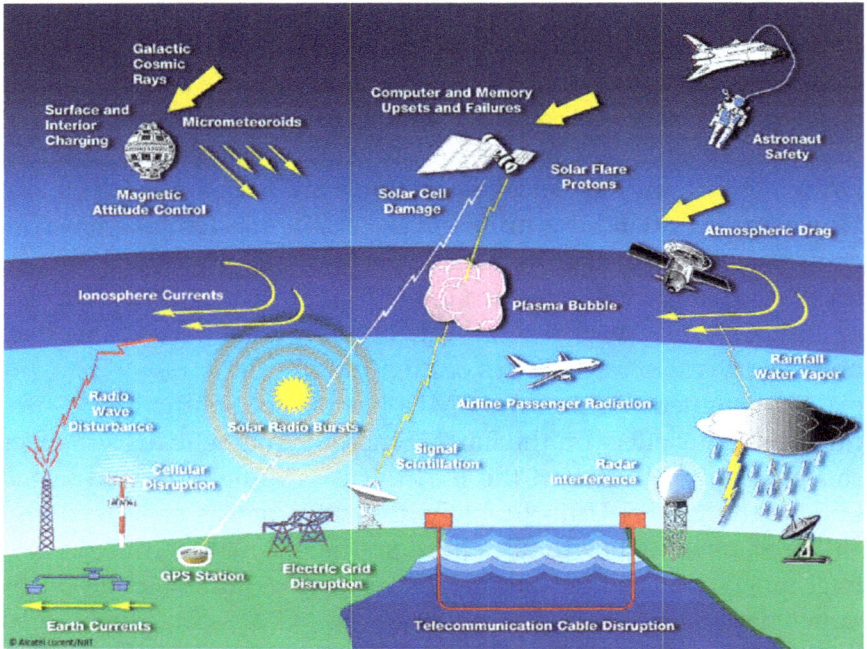

Figure 5.6.   Chart of the different effects of solar activity on Earth-bound activities. (Courtesy of NASA.)

5.4. High-Speed Streams

We have already seen that the solar wind does not flow uniformly but has high-speed streams embedded in a slower wind. These high-speed streams are also the places where the interplanetary magnetic field (IMF) changes polarity. They seem to emanate from coronal holes, which are dark regions in the corona that live for several months. These coronal holes, along with their associated high-speed streams, rotate with the Sun with a period of 27 days. Thus, the high-speed streams, along with the associated changes in the IMF, cross the Earth regularly every 27 days. A geomagnetic storm accompanies each crossing of a sector structure. Such storms that occur regularly every 27 days are called recurrent storms. Recurrent storms also create problems on the Earth, perhaps with less impact. The sector crossings can cause meteorological effects as well, i.e. direct changes in terrestrial weather. A definite change in the pattern of cyclonic activity follows each sector crossing. Sector crossings tend to suppress cyclonic activity. This effect does not appear to be severe for single crossings.

But if one considered the effect of a long period over which sector crossings occurred very regularly, then the accumulated anticyclonic effect might well lead to large periods of dry weather. Sector structures tend to be enhanced during periods of low solar activity. Major droughts in the North American plains also seem to have occurred during periods of unprecedented low solar activity. It is very tempting to relate the droughts to the sector structure.

5.5. Variability of Solar Luminosity

The solar radiative output is the result of energy generated in the solar interior by thermonuclear reactions. We have seen that these reactions are controlled very precisely, so that there is no scope for large variations in the solar luminosity. In addition, the energy transport mechanisms are so efficient that there is no stagnation of energy on its way outward. Thus, we expect the Sun's energy output to be very constant. But how constant is constant? A change of even a fraction of a percent in the solar luminosity can create a change of a few degrees in the Earth's surface temperature. This level of a sustained change over a long period is sufficient to cause either ice ages or global warming leading to the melting of the polar ice caps and consequent flooding by the rising ocean level.

Keeping such limits in mind, the solar luminosity is being monitored very precisely. And some strange results have emerged. It is now certain that the Sun decreased in luminosity from solar maximum to solar minimum by one part in a thousand. What does this mean in the long run? Does it mean that the number of sunspots somehow modifies the amount of light generated by the Sun? Having more sunspots cannot directly reduce the total light emitted by the Sun, because the diminished convective transport in sunspots must be compensated for by additional transport outside of sunspots — a very tiny change in the temperature gradient is sufficient for this. Therefore, we must look elsewhere for reasons to explain the change of the total solar luminosity in step with the sunspot cycle. Whatever be these reasons, it would mean that any unprecedented change in the level of solar activity would result in a corresponding change in the solar luminosity. The more obvious and large variations in the ultraviolet part of the sunlight due to solar activity do cause major changes in the ionic composition of the upper reaches of the Earth's atmosphere. So much so that the phenomenon called sudden stratospheric warming seems to occur during times of strong solar activity. And the different atmospheric layers are known to be so well coupled that stratospheric changes could affect the lower atmosphere, leading to changes in weather patterns. This is a topic which requires intensive study.

Long periods of very low activity existed in the past. If these periods were also long periods of a lower amount of sunshine, then one can perhaps link phenomena like the ice ages with solar activity. In the short term, we should be wary of tampering with the carbon dioxide content of our atmosphere, because the so-called "safe" limits for increase in these gases are based on a constant solar luminosity. If the Sun were to change by even very small amounts, the ecological consequences could be quite serious.

5.6. Solar Activity and Climate Change

There is a disquieting trend of views being bandied about, that the changes produced in the climate system can be explained by natural phenomena like solar activity. I call the trend disquieting because it tends to absolve humans from their sins of damaging the ecosystem and allows them to blame the Sun for climate change. This trend seems to be encouraged by vested interests selling fossil fuels in a big way.

What could be the connections between solar activity and climate? It is known that solar activity modulates the amount of cosmic rays reaching the Earth. Cosmic ray flux is smaller during the periods of the sunspot maximum, while it is more intense during periods of low activity. The cosmic rays produce a shower of charged particles when they strike the Earth's atmosphere. These charged particles can be centers for attracting water molecules, thus forming clouds. Indeed, some scientists have reported a good correlation between cosmic ray flux and cloud cover. Cloud cover can cause decrease in sunlight falling on the ground, thus producing a cooler atmosphere. This could explain why the periods of extremely low activity in the past coincided with periods of very cool climate. The opposite of this trend could then mean that periods of high solar activity should produce greater global warming. But we must be very careful about jumping to any conclusion. First of all, the relation between cloud cover and cosmic rays is not properly established, with many people contesting the claims for such a link. Second, even if there is a relation, the big question is: How much would this influence the global temperature? It is obvious from Figure 5.7 that the large increase in global temperature seen since the 1960s is not explicitly linked with solar activity.

Another point is about the depletion of ozone in our atmosphere by CFCs. All "safe" limits for the emission levels of these gases assume a basic ozone content that is more or less constant. However, the ozone "hole" does vary in step with the solar cycle. This is understandable, because the ultraviolet emission from the Sun,

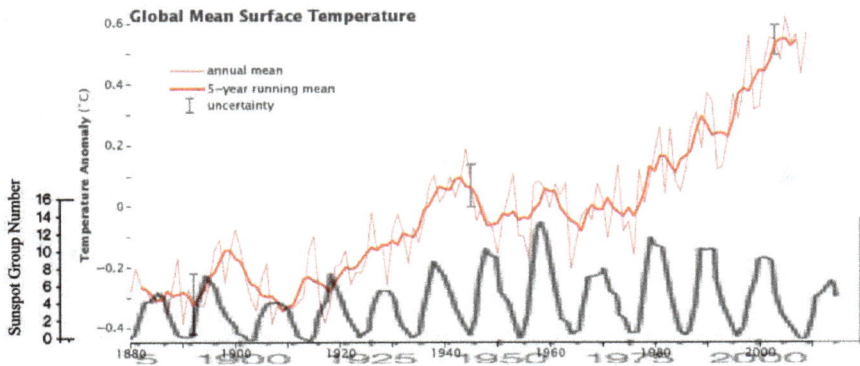

Figure 5.7. Comparison of solar activity measured in terms of the newly defined sunspot group number with that of global temperature. (*Source*: A combination of https://earthobservatory.nasa.gov/Features/GlobalWarming/images/giss_ temperature.png for temperature and http://physicsworld.com/cws/article/news/ 2015/aug/07/new-sunspot-analysis-shows-rising-global-temperatures-not-linked-to-solar-activity for sunspot group number. Composite created by the author.)

which produces the ozone layer in the first place, can change with the solar cycle. Knowing that our atmospheric chemical balance is delicately poised, it is foolish to advance "safe" limits for pollutants, simply because for all we know (and don't know), the Sun might suddenly increase its activity and pour out a high level of ultraviolet radiation. This extra radiation, combined with the existing levels of pollutants in the atmosphere, might lead to a spiraling chain reaction that might drastically alter the net ozone content. This, and similar other chilling possibilities, provide enough reasons for studying the Sun even more carefully than ever before. However, some recent studies have revealed that the Sun has spent only a few percent of the past 7000 years in such periods of high activity. Another study has shown that the solar activity has steadily declined over the past few cycles and the Sun may be heading toward a miniversion of the Maunder minimum. Fortunately, we now have better means of studying the Sun, and the next and final chapter of this book will describe these tools of the trade.

6 TOOLS OF THE TRADE

6.1. Introduction

I have tried to give you a flavor of the many details that have been collected over many years about our backyard star. However, the story is not complete without knowing how one can obtain these details. The Sun emits x-rays, ultraviolet rays, visible light, radio waves, and neutrinos. Let us look at the way we can detect and analyze each kind of emission. We will begin with the most common and well-known emission, namely optical radiation.

6.2. Solar Optical Telescopes

The special quality about solar observations is the fact that we can distinguish many tiny features on the face of the Sun, which we can hardly see on any other star. To exploit this fact fully, we must be able to form a large image of the Sun. A lens with a large focal length can form a large image. For example, to form a 10-centimeter image of the Sun, we need a lens that has a focal length of 10 meters. It would be extremely difficult to point at and track the Sun with any telescope that is so long. One might use a combination of lenses or mirrors to first form a smaller image, and then further magnifying it after folding the rays. This entails another problem — that of the heating produced by focussing sunlight with a short focus. Astronomers found a clever way out. They rotated a plane mirror about an axis that is parallel to the Earth's axis of rotation. One can then tilt the mirror at a suitable angle to the Earth's equator, so that the sunlight, after reflection at the mirror, travels along a direction parallel to the polar axis. We can form an image as large as we like, by having a long-enough focal length for the lens. Thus, we can obtain a large image without moving a long telescope. Such a mirror is known as a heliostat. An annoying problem with a heliostat is that the image rotates during the course of the day. The solar image formed with a heliostat will rotate by 180 degrees during the Sun's movement from sunrise to sunset. The astronomer Foucalt invented a variant of the heliostat in 1869, which he called the siderostat. In this system, there is an additional movement of the rotating mirror to keep the reflected beam of sunlight always in a horizontal direction.

Gabriel Lipmann, who won the Nobel Prize in 1908 for his invention of color photography, solved the problem of image rotation in the following way. He also used a mirror that rotated about an axis parallel to the Earth's axis. However, the mirror was perpendicular to the plane of the Earth's equator. In addition, the speed of rotation of the mirror was half of the Earth's speed. In this way, reflected sunlight traveled along a constant direction, during any particular day. Another mirror received this reflected sunlight, and in turn sent this light along a convenient direction, using a suitable tilt. The image formed by such a two-mirror system, called a coelostat, did not rotate during the day. The position of the second mirror relative to the first mirror had to change from day to day, as the Sun moved from north in summer to south in winter. Professional astronomers like to place their coelostats tens of meters above the ground surface. This is because heating of the ground during the day produces turbulence, which can degrade the quality of the images. At such heights, a large portion of this turbulence is below the coelostat, so that a clear view of the Sun is possible. Figure 6.1 shows an example of a solar telescope that uses a coelostat.

Professional astronomers also choose the location of their telescopes very carefully. The crispest images are possible when the observatory is in the middle of a large lake. The air above the lake does not heat up during daytime and thus does not disturb the quality of the images. Figure 6.2 shows the Udaipur Solar Observatory, with an example of a modern solar telescope, the Multi

Figure 6.1. A 60-centimeter coelostat of the Kodaikanal Observatory seen through the slit of the dome housing the Kodaikanal Tower/Tunnel Telescope (KTT). (Courtesy of Indian Institute of Astrophysics.)

Application Solar Telescope (MAST), situated on an island in Lake Fatehsagar in Udaipur, in the desert state of Rajasthan in North India.

A high mountain site is also preferred because the air is clear and dustfree. Under such ideal conditions, it is even possible to see the coronal emission without a total solar eclipse, by blocking out the central portion of the solar image and reimaging the rest of the coronal light. Figure 6.3 shows the solar tower/tunnel telescope at the Kodaikanal Observatory as an example of such an observatory located 2300 meters above mean sea level atop the Palani Hills in South India.

One can analyze the light from the Sun in a variety of ways. The total amount of sunlight is carefully measured to look for long-term trends in the solar luminosity. Very often, a small portion of the solar image is sent through the entrance slit of a spectrograph to get the spectrum of the sunlight. We generally see some dark lines at specific locations in the solar spectrum along with a continuous

Figure 6.2. *Top*: View of the Udaipur Solar Observatory, on an island in the middle of Lake Fatehsagar of Udaipur. The shorter building on the right houses the newly commissioned Multi Application Solar Telescope, covered by a collapsible dome made of a special fabric similar to the fabric used for the tops of Ferraris. *Bottom*: View of the telescope with the dome completely folded down and with the back of the 50-centimeter aperture primary mirror in the foreground. (Courtesy of Udaipur Solar Observatory/Physical Research Laboratory.)

Figure 6.3. The solar telescope of the Indian Institute of Astrophysics (IIA) located 2300 meters above mean sea level atop the Palani Hills in South India. (Courtesy of IIA.)

background spectrum. These dark lines occur wherever atoms of particular elements constituting the solar atmosphere absorb a lot of light at a particular wavelength. Each atom absorbs a specific set of wavelengths in a specific intensity pattern. A careful study of this pattern gives you the relative number of atoms of each element present in the solar atmosphere.

Sometimes, the wavelengths of these lines shift by a small amount by the Doppler effect. Using the Doppler effect, we can measure the speed of movement of the solar plasma. The five-minute oscillations were first detected from such measurements. A magnetic field also has a distinctive effect on the shape of spectral lines. Detection of the magnetic field in sunspots made use of this effect (the Zeeman effect). Here again, a spectrograph disperses the light and provides the means to look at the shape of the Zeeman-distorted spectral lines.

Figure 6.4.   View of the adaptive optics system being developed at the Udaipur Solar Observatory. (Courtesy of USO/PRL.)

A serious problem in observations using visible light is the fact that the turbulence in the Earth's atmosphere can severely degrade the image quality. You will see such a distortion of images when you look at objects across a paved road on a hot day. One way of restoring the image quality is to have an adaptive optics system. This consists of a flexible mirror surface that will be deformed to match the wave front of the incoming light which is distorted by the atmosphere. When light is reflected off this distorted mirror surface, the resulting wave front becomes straight again and produces an undistorted image! This is a very challenging task which requires the orchestration of the surface of a deformable mirror controlled by computers that calculate the amount of image distortions at speeds of 1000 cycles per second and then give commands to deform the deformable mirror by the required amount.

Interestingly, many of the working models of adaptive optics systems have resulted from research done for the Star Wars program, technically known as the Strategic Defense Initiative (SDI). The SDI basically was designed to knock down enemy missiles using powerful lasers. One severe problem with lasers is that the beam gets broken up into laser speckles while passing through the turbulence of the Earth's atmosphere. So the SDI created a system of deformable mirrors which could compensate for the distortions created by the Earth's atmosphere. These were one of the first successful adaptive optics systems created. This example of astronomy benefiting from military applications is not new or unique, however. The very first telescopes of the 17th century detected the movements of enemy troops or ships from a long distance. The spinoffs gained from war efforts have examples in radio as well as x-ray astronomy.

6.3. Radio Telescopes

The discovery of radio emission from the Sun was purely a matter of chance. A very large sunspot appeared on the face of the Sun in February 1942. At the very same time, British radars were carefully monitoring the sky for signs of enemy aircraft during the course of the Second World War. Radar operators could not locate the source of a strong radio interference that jammed their radar sets and hindered their surveillance. This became a matter of great concern for a while, until scientists tracked down the culprit, which was the Sun. After the end of the War, this discovery was published, and the science of solar radio astronomy was born. Thereafter, scientists all over the world started systematic studies of the radio emissions from the Sun. The first of the radio telescopes was in the form of a radar dish antenna. The radio waves were concentrated on to the focus of the dish. A hornlike feed collected the radio signals and transmitted them to suitable radio receivers.

Those of you who are familiar with radio antennas know that the antenna does not respond equally well to radio waves coming from all directions. There is a definite pattern associated with each antenna's response. For example, the direction of maximum response would be perpendicular to the face of the dish. The response is strong only for a cone of angles surrounding the central direction of maximum response. The angular size of this cone is smaller when the dish is larger. Thus, we would require a dish more than 100 meters in diameter to image the Sun, while, to be able to see sunspots, we will need a dish more than 3 kilometers in diameter.

Clearly, it is a formidable task to erect such a huge dish and make it track the Sun. Fortunately, radio astronomers have found a clever way to image the Sun and other celestial objects using interferometry. In this technique, the radio waves detected at a large number of independent radio dishes are combined to produce an "image" with the required spatial resolution. The larger the separation of the individual dishes, the better the resolution. Figure 6.5 shows one of the dishes of the Giant Metre Radio Telescope located near Pune in India. The turbulence in the ionosphere affects lower-frequency waves. All frequencies must be

Figure 6.5. One of the several radio dishes that form the Giant Metre Radio Telescope of the National Centre for Radio Astronomy, situated near Pune, India. Though meant for studying external galaxies, this telescope has also been used to map active regions and flares. (Courtesy of NCRA/TIFR.)

Figure 6.6. View of one of the arms of the array of dipoles that make up the Gauribidanur Solar Telescope, located 50 kilometers northeast of Bangalore, India. (Courtesy of Indian Institute of Astrophysics.)

protected from interference by man-made radio signals used in telecommunications and television broadcasts. Countries enter into treaties from time to time to reserve a set of frequencies for the astronomers, but the demand from providers of telecommunications services has been always increasing. As mentioned earlier, the Gauribidanur Radio Telescope is one of the few radio telescopes operating at the very low frequencies that come from the outer solar corona (Figure 6.6).

There is yet another different kind of radio telescope whose antenna is shaped like a long cylindrical parabola with the axis of the cylinder pointing toward the pole star. The cylinder is made to rotate about its axis at a speed of one rotation per day. In this way, it can track any object in the sky. By means of electronic switching, the beam of the antenna can be made to scan over different celestial latitudes (declination, as called by astronomers). Such a telescope, belonging to the Radio Astronomy Centre of the Tata

Figure 6.7.   The OOTY radio telescope. (Courtesy of NCRA/TIFR.)

Institute of Fundamental Research and situated at OOTY in South India (Figure 6.7), has been used in recent times to map the density and speed of the solar wind over a wide range of distances from the Sun.

6.4. X-Ray Telescopes

You have already learnt that the corona is hot enough to emit x-rays. x-rays and ultraviolet rays from the Sun cannot reach the surface of the Earth, because they are completely absorbed by the Earth's atmosphere. Astronomers had no choice other than to try to put their detectors outside the Earth's atmosphere, in outer space. Here again, the Second World War indirectly contributed to astronomy. The US army captured a large quantity of the famous V-2 rockets and brought them back to the US. Initial successes were obtained in the initial scientific experiments, around 1946, using these rockets. The warhead of the rocket was capable of holding 900 kilograms of explosives. A scientific instrument replaced the warhead. The rockets reached heights of 120 kilometers above the ground and remained in the air for about 6 minutes. However, they thereafter crashed down at a speed of more than 3000 kilometers per hour, so some methods were devised to slow down the falling rocket. Parachutes then brought down the instruments in a gentler way. Initially, these rockets measured the amount of x-rays and ultraviolet rays emitted by the Sun at different wavelengths. The Sun emits extra energy in a few spectral lines in x-rays and the ultraviolet. This is opposite to the case of visible light from the photosphere, which is mainly a broad continuous emission interspersed with absorption lines. The earlier rocket flights used beads of a material called lithium fluoride arranged in front of a spectrograph. These beads produced many different images of the Sun in the ultraviolet emission lines. The early images were not very sharp, because the rockets were not able to point steadily at the Sun. Later on, the scientists succeeded in making pointing systems that produced better images. When satellites became available, astronomers went ahead and used such satellites as platforms to carry their telescopes.

The modern x-ray telescopes based on such satellites can pick out the x-ray emission from a solar flare that is only a few seconds of arc in angular size. Since x-rays travel in straight lines through most materials, it is difficult to bend these rays. The way of forming proper x-ray images depends on the energy of the x-ray photon. Very-high-energy photons are called hard x-rays and their wavelengths are of the size of single atoms. Lower-energy photons, called soft x-rays, have wavelengths that are about ten to a hundred times larger than atoms. Early versions of soft x-ray telescopes used Bragg reflections at grazing incidence. In modern versions, a stack of thin coatings of certain metals is deposited one over the other. Here normal incidence is used. It is even more difficult to image hard x-rays because of their smaller size. For this, a technique very similar to that used in a pinhole camera is employed. Several pinholes or slits are placed in the path of the hard x-rays. The diffraction of x-rays is used to reconstruct the images. Here again, as in the case of radio waves, the images are reconstructed after huge mathematical calculations using computers (Figure 6.8).

The high-energy physics of solar flares continues to be of interest to solar physicists. Thus, even information from the unresolved solar disk is very useful. Such a payload was sent piggyback on

Figure 6.8. An Artist's concept of the satellite *RHESSI* carrying a spectroscopic imager payload. (*Source*: https://hesperia.gsfc.nasa.gov/hessi/images/hessi craft.gif)

GSAT-2 on May 8, 2003, and got some new results on solar flares (Figure 6.9).

6.5. Neutrino Detectors

You have already encountered the facts about the tiny neutrinos that directly escape from the solar interior and shoot out into space. These neutrinos interact only weakly with matter and are therefore very difficult to detect. This did not stop the scientists from devising ingenious ways of measuring the number of neutrinos emitted by the Sun per second. The scene of operations this time is

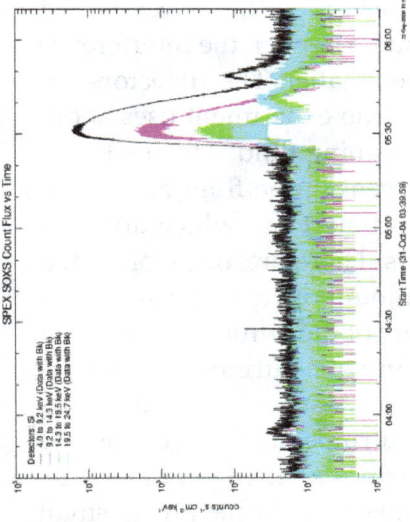

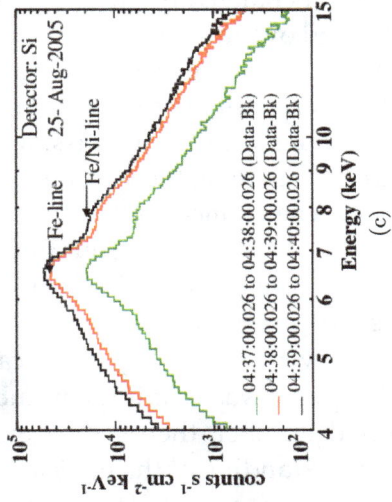

Figure 6.9. The SOXS payload (*middle*) riding piggyback on GSAT-2 (*left*), with a sample spectrum taken during a solar flare. (Courtesy of Dr. Rajmal Jain, PI of SOXS payload.)

neither in space nor on mountaintops, but deep within the Earth. Deep underground mines are ideal places to locate the neutrino detectors. This is to shield the detectors from the cosmic rays and other stray particles that would have swamped the detectors had they been placed on the ground. As you go deeper, the interference from these unwanted particles will be smaller. The detectors are usually large quantities of pure fluids. One experiment uses about 500 tonnes of perchloroethane, a dry-cleaning fluid. The neutrinos interact with the nuclei of chlorine present in the fluid and transform these into argon nuclei. Argon is an inert gas, which does not chemically combine with other elements. Therefore, once formed, it remains unchanged in the fluid. The amount of argon accumulated in the fluid over a period gives a measure of the number of captured neutrinos. From the probability of capture of neutrinos by chlorine, we can then estimate the number of neutrinos incident on the tank. By assuming a uniform emission of neutrinos in all directions, we can finally estimate the number of neutrinos emitted per second by the Sun. The chances of neutrino capture by chlorine are so small that scientists expected about three detections in two days from the entire 500 tonnes of fluid (Figure 6.10). What startled both particle and solar physicists was the fact that the actual number detected was one event in two days. This caused a general panic. The reason for this panic was that if the number detected is only a third of the expected number, then something must be seriously wrong with our understanding of the nuclear reactions in the solar interior. And the entire subject of stellar physics rests happily on the assumption that we really know the source of energy generation within stellar interiors.

Yet another attempt was made to measure the solar neutrino flux in order to detect neutrinos with a different energy range, closer to the energy of the neutrinos produced by the main thermonuclear reaction, in contrast with the Homestake experiment, which could detect neutrinos only from a side reaction. This second experiment (Figure 6.11) used the Super-Kamiokande detector, which was originally designed to search for proton decay, study solar and atmospheric neutrinos, and keep watch for supernovae in

Figure 6.10. Picture of the chlorine neutrino experiment at the Homestake Mine, South Dakota, USA, where the first neutrino detector was built. (Courtesy of Nick Strobel at www.astronomynotes.com.)

the Milky Way. A neutrino interaction with the electrons or nuclei of water can produce a charged particle that moves faster than the speed of light in water (although of course slower than the speed of light in vacuum). This creates a cone of light known as Cherenkov

Figure 6.11.   The Super-Kamiokande, or Super-K for short, is a neutrino observatory in the city of Hida, Gifu Prefecture, Japan. [Credit: Kamioka Observatory, ICRR (Institute for Cosmic Ray Research), University of Tokyo.]

radiation, which is the optical equivalent to a sonic boom. The Cherenkov light is projected as a ring on the wall of the detector and recorded by several thousands of photomultiplier tubes (PMTs). By the use of the timing and charge information recorded by each PMT, the interaction vertex, ring direction, and flavor of the incoming neutrino are determined. This was invented by Masatoshi Koshiba of the University of Tokyo, Japan, who shared the 2002 Physics Nobel Prize with Raymond Davis and Ricardo Giacconi. Alas, this experiment could not detect the required number of neutrinos either.

The clinching evidence to resolve the mystery of the missing neutrinos came with the measurement of all the three kinds of neutrinos by the Sudbury Neutrino Observatory (Figure 6.12). This was a heavy-water Cherenkov detector designed to detect neutrinos produced by fusion reactions in the Sun. It used 1000 tonnes of heavy water loaned from Atomic Energy of Canada Limited (AECL), and contained by a 12-meter diameter acrylic

Figure 6.12. The Sudbury Neutrino Observatory of the Queen's University (Canada), Lawrence Berkeley National Laboratory (USA), University of Pennsylvania (USA), University of Washington (USA), Oxford University (UK), Los Alamos National Laboratory (USA), University of British Columbia (Canada), Carleton University (Canada), University of Guelph (Canada), Laurentian University (Canada), Brookhaven National Laboratory (USA), University of Texas at Austin (USA), LIP–Lisboa (Portugal), and Massachusetts Institute of Technology (USA). (http://www2.lbl.gov/Science-Articles/Archive/sudbury/sno3. html.) (Courtesy of Lawrence Berkeley National Laboratory. Photographed by Roy Kaltschmidt.)

vessel. Neutrinos reacted with the heavy water to produce flashes of light called Cherenkov radiation. This light was then detected by an array of 9600 PMTs mounted on a geodesic support structure surrounding the heavy-water vessel. The detector was immersed in light (normal) water within a 30-meter barrel-shaped cavity (the size of a 10-story building!) excavated from Norite rock. Located in the deepest part of the mine, near Ontario, Canada, the over-burden of rock shielded the detector from cosmic rays. The heavy water was borrowed from the Canadian Atomic Energy Agency, and was promptly returned to the agency after collecting enough evidence about the solar neutrinos. After a lot of patient work, a glimmer of truth has appeared. Apparently, the neutrinos flip from one form of neutrinos to two other forms of neutrinos that cannot be detected by the earlier experiments. In the laboratory, these changes occur with insignificant probability, but these transformations are amplified when neutrinos pass through matter. Since the solar neutrinos have to traverse a large amount of matter before they come out of the Sun, it is quite possible that the transformations proceed drastically enough to cause a deficiency in the number of neutrinos of the right kind actually reaching the earlier detectors. Since the SNO measured all the three known kinds of neutrinos, the total number detected exactly matched the number predicted from the fusion reactions happening in the solar interior. Physicists and astronomers had to work together to arrive at this conclusion.

The solar astronomers employed all their skills to make sure that the internal properties of the Sun were measured accurately using sound waves to probe the solar interior. The particle physicists employed all their knowledge about the fundamental forces of nature to figure out the solution to the neutrino problem. By a combination of hard work and clever thinking, these two groups of scientists have arrived at a new understanding of matter. The neutrino oscillation from one type of neutrino to another was definitively confirmed by the Sudbury experiment. Thus, we see that many people came together to solve the serious riddle about the missing solar neutrinos once and for all. And ended up with revolutionary evidence of neutrino oscillations!

We have come to the end of this story about the Sun and the efforts that were put in to understand the workings of our daytime star. Despite all these efforts, there are still some pesky unknown things about the Sun. Mainly, we still do not know how the solar cycle is maintained. We still do not know how the corona is heated.

We still do not know how the solar wind gets accelerated. We still do not know how to predict solar activity. And we still do not know whether solar activity modulates our terrestrial weather and climate. So, the new generation of young bravehearts must continue to struggle and find out the answers.

SUGGESTED READING

Fundamentals of Solar Astronomy, by A. Bhatnagar and William Charles Livingston (World Scientific, 2005).

The Sun Kings, by Stuart Clarke (Princeton University Press, 2009).

Nature's Third Cycle: A Story of Sunspots, by Arnab Rai Choudhuri (Oxford University Press, 2015).

Mr. Tompkins in Wonderland, by George Gamow, (The Macmillan Company, 1946).

INDEX